Why the Wind Blows

Why the Wind Blows

A History of Weather and Global Warming

MATTHYS LEVY

Illustrations by
Sue Storey

Upper Access, Inc., Book Publishers

Published by Upper Access, Inc., Book Publishers
87 Upper Access Road, Hinesburg, VT 05461
802-482-2988 • www.upperaccess.com

Cover design and interior layout by Kitty Werner, RSBPress LLC
Illustrations by Sue Storey

Polar bear photo © Martin Will–Fotolia.com
Storm photo © Kitty Werner
Back cover glacier photo © Stephan Hoerold–iStockPhoto.com
Author's photo by Kitty Werner

ISBN: 978-0-942679-31-1
(ISBN 10: 0-942679-31-8)

Library of Congress Cataloging-in-Publication Data

Levy, Matthys.
 Why the wind blows : a history of weather and global warming / Matthys Levy ; illustrations by Sue Storey.
 p. cm.
 ISBN 978-0-942679-31-1 (alk. paper)
 1. Weather. 2. Global warming. I. Title.
 QC981.L499 2007
 551.5--dc22
    2006038756

Printed on recycled acid-free paper in the United States of America

07 1 2 3 4 5 6 7 8 9 10

Contents

Acknowledgments

The critical review and valuable suggestions provided by Dr. Margaret A. LeMone of the National Center for Atmospheric Research and Dr. Alan K. Betts of Atmospheric Research helped in the development of what were scattered thoughts. The manuscript evolved over a period of years and was also read by colleagues and friends too numerous to name who gave me comments that I found invaluable in reshaping the work. I am also indebted to my publisher, Steve Carlson, who undertook this work with enthusiasm and to Kitty Werner, who led me to the publisher and prepared the manuscript for publication. Finally, I am grateful to my wife, Julie, for her support through the long process that finally led to the publication.

Introduction

All the fountains of the great abyss burst
 forth,
and the floodgates of the sky were opened.

Genesis 7:11

After having written, with my mentor and friend, the late Mario Salvadori, stories of why structures fail and the causes and consequences of earthquakes and volcanoes, I felt the need to continue exploring the origins and effect of natural forces. The most powerful of all is the force that gives us life, warms our planet, bathes us in light, *and* causes all weather related disasters: the sun's radiation.

Every day, somewhere on Earth it rains, perhaps in a torrent; the wind blows, perhaps with the force of a hurricane or typhoon; and it snows; perhaps as a blizzard. As a youngster, I spent one summer camping on an island in the middle of a lake when the sky turned grey and stormy. For days, we were drenched by rain and the wind blew, raising whitecaps on the normally calm surface of the lake, preventing us from taking our canoe to reach the shore to restock our dwindling food supplies. Naively, I thought we would never escape our island prison and, more importantly, that I would starve to death. My father reassured me that neither would happen and after three days the storm abated and I was saved. Since then, I have experienced the fury of a North Pacific storm while on a troopship, blinding Northeast snowstorms, and devastating flooding, but all were mild compared to the destruction wrought by God in the biblical story of Noah.

Before unleashing forty days and nights of rain, as punishment for man's wickedness, God commanded Noah to build an ark to save selected species of animals and birds. The ark was to be 134 m (440 ft) long, 22 m (73 ft) wide and 13 m (44 ft) high, and have three decks. Noah had seven days to complete the task (quite a job for a 600-year-old man) before the floods came and covered the highest mountains, wiping out all living things except those safely ensconced on the ark. The story was apparently derived from a Mesopotamian tale of a great flood preserved in the eleventh tablet of the Gilgamesh Epic with one difference: That ark was a 54 m (176 ft) cube.

The story fails to take into account a physical reality. Of all the water on Earth, 97% is already in the oceans and 2% is in the form of ice, mainly in the polar caps. The remaining 1% is in all the lakes, rivers, underground water, and in the air as water vapor. If all of today's glaciers were to melt, the sea level would rise 70 m (230 ft). It is therefore impossible for the entire world's water to cover the highest mountains. While the story is certainly apocryphal, it is nevertheless symbolic of the devastation that can be caused by natural forces in a particular area. The limited "world" of the story's writer could indeed have been destroyed by a flood, as the areas of every great river basin are, even today, periodically flooded. It is known that the Black Sea was once a fresh-water lake, about two thirds of its present size. About seventy-five hundred years ago, after melting glaciers had raised sea levels following the end of the last ice age, the Mediterranean Sea breached the Bosporus Valley. Salt water poured in from the Sea of Marmara, raising the level of the Black Sea by almost 180 m (600 ft) and extending its boundaries. As the water level rose about 150 mm (6 in) per day, the inhabitants along the shore of the Black Sea would have viewed the event as a great flood, perhaps giving rise to the biblical tale.

The story of weather is totally intertwined with the story of humankind. About forty-two hundred years ago, the cradle of civilization, in the verdant Garden of Eden in the fertile crescent between the Tigris and Euphrates rivers, was devastated by a three-hundred-year-long drought. A dry, barren landscape replaced the lush fertile valley and destroyed the thriving agricultural society that had developed there. It was all part of a long-term cooling and drying cycle that still affects us today. In ancient Mesopotamia, not only were there gods of war (Ishtar) and water (Ea) among others, but weather was recognized with its own god, Adad.

The Chinese had observed sunspots, those dark seas on the face of the sun that indicate solar magnetic storms, before the first millennium. Over time, intense sunspot activity had been found to follow an eleven-year cycle. However, suddenly during the 1645–1715 period, sunspot activity decreased dramatically, virtually coincident with a period of intense cold in the Northern Hemisphere that was called the Little Ice Age. This and many more examples demonstrate the influence of weather on human activity.

A little over a century ago, there began an ominous trend that continues to this day whereby human activity began to influence future climate. Man had always tried to control the consequences of weather phenomena—building shelters to provide a shield from rain, snow, and extremes of heat and cold; conduits to channel rainwater; and breakwaters to calm ocean waves and create a safe port. There have also been attempts at active manipulation of weather, such as by cloud seeding to create rain. Passive changes took place starting in the late nineteenth century, fueled by rapid industrialization, an expanding use of carbon-based fuels, and a booming population. Climate change is now in the throes of global warming that can no longer be attributed to "natural" long-term trends. Although changes that can be ascribed to global warming arrived gradually at first, that may no longer be true in the twenty-first century, as significant and possibly abrupt changes in climate may be expected on our planet.

The evolution of the understanding of climate and weather changes is explored in this book by relating adventures of those who lived through it. It is a story that can be seen from the macro-view of long-term trends, such as the ice age cycles, to the micro-climates over your home town.

And it all starts with the sun!

1

Imported From Siberia

Meteorology and the Origins of Weather

Sometimes I go about in pity for myself
and all the while,
A great wing carries me across the sky.

Anon. Ojibwa Tribe

The sun, as it breaks the horizon in the early morning, sends us its warming rays, giving us the energy to venture forth and engage another day. Those same rays warm the earth and the oceans, raising their surface temperature, causing moisture from the seas to evaporate and rise up into the atmosphere to form clouds and, in the process, set into motion a local circulation that gives rise to winds. But the story of where the wind comes from is not so simple, for three reasons.

First, the sun is not a perfect radiator and does not send us a uniform amount of radiation. Its surface is spotted with cooler regions first observed by Chinese astronomers in the first century BC.

These appear as dark spots, which become more intense about every eleven years and are associated with an increase in the sun's radiation. As a result of these **sunspots,** the difference in the amount of energy sent to Earth (as well as all the other planets in our solar system) is about two tenths of a percent. This is seemingly small, but causes interference in radio signals and is enough to lead some observers to claim an increase in agricultural yields as a result of increased temperatures.

A second influence in global weather arises from the fact that the Earth travels around the sun, as first noted by **Aristarchus of Samos** 2,300 years ago but suppressed during centuries of Earth-centered beliefs until bravely rediscovered by **Copernicus** in the sixteenth century. Although first thought to be circular, the path taken by Earth in its travels through the heavens was shown by **Johann Kepler** in 1609 to be described by an ellipse with the sun as one focus (Fig. 1.1). As it undertakes its year-long voyage around the sun, Earth passes closest to it, at 146 million km (91 million miles) around July 4th and is farthest, at 151 million km (94 million miles), around Jan 3rd.[1]

However, the most significant influence in causing non-uniform radiation from the sun comes not because of its elliptical path but from the fact that the Earth rotates, alternately warming and cooling the surface as we pass from day, facing the sun, to night, facing the icy universe. Furthermore, the rotation takes place about

---

1   The path of the Earth around the sun changes slightly over millennia so the distances cited here are the current ones.

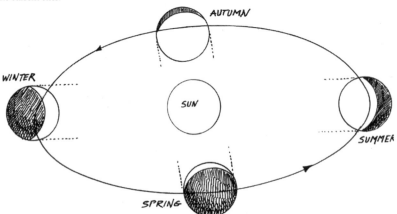

Fig. 1.1 Earth's voyage around the Sun

an axis that is tilted by about 23.5° from the *ecliptic plane*, a plane defined by the Earth's path around the sun. This tilt is the reason we experience four seasons. In the Northern Hemisphere, we are tilted farthest away from the sun on December 21st or 22nd (the **Winter Solstice**) as the sun is directly over the **Tropic of Capricorn,** even though two weeks later, our orbital path around the sun brings us closest to it. On June 21st or 22nd (the **Summer Solstice**), the sun is directly overhead the **Tropic of Cancer,** and the Northern Hemisphere is tilted closest to it even though two weeks later, the Earth's orbital path takes it farthest from the sun. The difference in the dates for the solstices arises from the fact that the Earth takes about 365 days, 5 hours, 48 minutes, 46 seconds to complete its orbit. Since we cannot deal with fractions of days, leap years, with an additional day, have been introduced every four years. To further correct the overcompensation caused by the establishment of leap years, leap centuries were introduced, in which a leap year is skipped, except every 400 years, to adjust for the millennium correction.

Since we, in the Northern Hemisphere, are warmer in summer than in winter, it is clear that the Earth's tilt is more significant in making our weather than is its distance from the sun to the earth. In between these two extremes are the spring and fall **equinoxes** on March 21st or 22nd and September 22nd or 23rd when the sun is directly overhead the equator and night and day are about equal.

The warmth of the sun is fueled by the continuous thermonuclear fusion reaction taking place deep within its core. It is a process in which two atoms join to create a new atom and, as Einstein showed, release lots of energy. The sun's warmth is most intensely felt around the equator, the region that is heated most intensely every day and is almost directly facing the sun. This is why the equatorial regions are hot all year. On the other hand, the polar regions, which experience longer days, receive only a glancing blow of the sun's radiant energy since they face out toward space and not toward the sun (Fig. 1.1). As a result, the poles tend to remain cold and experience extremes of weather between summer and winter.

Travels Through Space

The surface of the sun is an incredibly hot 6 000°C (11,000°F). As we face the sun on a clear beach day, we can feel its warmth and see its light, but how does this energy reach us? There are three ways in which energy can be transmitted: **conduction**, **convection** and **radiation**. Conduction is a direct means of transfer in which vibrating particles collide with each other but rely on a conducting medium such as is present in an aluminum or copper pot, both of which conduct heat from the flame on the stove to the pot's contents. The near vacuum of space is certainly not such a medium, and even air is an insulator rather than a conductor. The insulating property of air is used, for instance, in insulating windows where two sheets or glass are separated by a trapped layer of air that in winter holds in the warmth of the room while in summer keeps the heat outdoors.

Convection is the movement of currents in a gas or liquid such as those seen to travel as bubbles in a pot of boiling water. According to **Archimedes's Principle**, the warmer (less dense) parcels tend to rise through the cooler surrounding medium (liquid or gas) setting up a convective current (Fig. 1.2). We will see later how important these convective currents are on Earth in the development of weather, but the void of outer space, lacking particles of matter, does not provide an environment conducive to convection.

Lastly, radiation from a source such as the sun involves the motion of waves that can travel in any medium, even the vacuum of space. The visible light emanating from the sun is one form of its radiant energy that reaches us, but there are many others, including infrared, ultraviolet, X-rays and microwaves, all classed as **electromagnetic waves** that move silently and invisibly through the cosmos.

Fig. 1.2 Convection

Of all the sun's energy that reaches Earth, only about half strikes the surface and is absorbed by the land and the sea, while the balance is either reflected by the atmosphere or absorbed within it. Thick, rich forests and bodies of water absorb almost 90% of the radiation that gets through the atmosphere, while snow and ice reflect as much as 80%, which is another reason why arctic regions are cold and remain cold even in direct sunlight, while tropical forests are hot and remain hot. Remember that air is compressible and that cold air is more compressed—and therefore heavier—than hot air, so it tends to remain on the surface while hot air, because it is lighter, rises. As a consequence, above the tropics, the air that is constantly heated is hot, while above the poles, the air remains cold for lack of solar radiation.

Equilibrium

Fortunately for us, the Earth is covered by air that fuels our bodies and feeds our plants: a gas made up of 21% Oxygen, 78% Nitrogen, and 1% other gases. The farther we are from the Earth's surface the thinner the air becomes. Air pressure decreases because of a combination of the decrease in the weight of the air column remaining above it, and, to a lesser extent, a decrease in the attractive force of gravity. When we pass above the **exosphere** (about 1280 km [800 mi] from the surface of the earth) there are no more air particles (molecules), and we reach the boundary of outer space. Our exploration of the origins of wind and weather takes place below the stratopause, about 50 km (30 mi) from the Earth's surface.

Weather is governed by one overriding principle: the *movement toward equilibrium*. Hot air, because it is lighter, rises toward colder altitudes and is replaced by colder air moving in, and cold air tends to gravitate toward hotter regions. Each of these movements illustrates the tendency of bodies of air to try to find a moderating temperature, not too hot and not too cold. As we have already seen, the air above the equatorial region is hot and that above the polar regions is cold. These two opposite air masses, not separated by a physical boundary, cannot long exist without moving toward equilibrium.

Looking at a slice of the atmosphere starting at the equator, hot air rises and moves north and south toward the polar regions. About one third of the way to the poles, it cools sufficiently to drop

back to Earth and split into two branches. One moves back toward
the equator along the surface, picking up heat in the process and
beginning a convective circulation called the **Hadley cell,** named
after George Hadley, an eighteenth-century British lawyer who first
identified it. The second branch moves along the surface toward
the poles, establishing a second convective circulation called the
mid-latitude cell (Fig. 1.3). This second circulatory pattern picks
up enough heat along its way north (or south in the Southern Hemisphere) to rise up as it reaches a position about one third of the way
closer to the poles. This gives rise to one last convective circulation,
the **polar cell,** caused by the cold polar winds being warmed as they
move away from the poles and rising about one third of the way to
the equator. Looking at a cross section of the Earth, these three
convective circulation packages appear as counter-rotating cells in
a north-south plane. But wait a moment! The Earth's rotation dis-
turbs this neat picture.

Looking down from the North Pole, the Earth rotates in a
counterclockwise direction. Therefore winds, seen from the van-

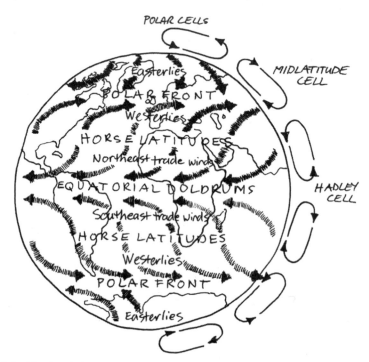

Fig. 1.3 Air circulation patterns around the earth

tage point of space to be moving in a straight line from the poles to the equator, actually follow a curved path toward the west on the Earth's surface. This effect, first attributed to the Earth's rotation by George Hadley in 1735, was described mathematically by Gaspard Gustave de Coriolis, a French scientist after whom it is named the **Coriolis Effect.** As a result of the Coriolis Effect, the three convective cells are distorted from their north-south orientation and create winds at the surface. As we move from the equator to the poles, these winds are called the trade winds, the westerlies[2] and the easterlies because of their prevailing surface wind directions (Fig. 1.4).

Pressure

Air moves about constantly, fueled by the necessity to find a mean temperature. But there is another characteristic of air that results in its movement: **pressure**. Air is a gas which consists of molecules that move in all directions, bouncing off each other and, like the push felt when standing under a shower, pressing against anything placed in their path. That pressure depends on how many molecules there are and how fast they are moving, just as the pres-

---

2   Wind direction is indicated based upon wherefrom it blows.

Fig. 1.4 Coriolis Effect

sure felt in a shower depends on the size and intensity of the spray.[3] When the temperature is high, the molecules are moving faster, resulting in higher pressure than when the air is cold and the molecules are moving more slowly. Near outer space, where there are fewer molecules, the pressure is low compared to what it is on the Earth's surface, where there are more molecules in a given space.

On the Earth's surface another phenomenon contributes to increased pressure: The force of gravity squeezes air molecules together under the weight of all the air above. Anyone who has flown up in an airplane or ridden a fast elevator has felt the consequences of this change of pressure with height. On the way up, decreasing air pressure sucks out the eardrum until air flows around it through the Eustachian tube to reinstate the balance of air pressure on both sides of the eardrum. The reverse occurs in a descent as pressure builds against the eardrums. The changes in air pressure felt in our joints and sinuses, as well as our ears, cause some people to say they can "feel" approaching storms (with their resulting drop in air pressure).

3   The change in wind per unit time equals the pressure gradient force plus the Coriolis acceleration plus the effect of turbulent mixing plus the effect of buoyancy (which accelerates the wind in a vertical direction only) plus the acceleration when the motion is curved (centrifugal force). In the case of the shower, the pressure gradient equals the shower pressure.

Fig. 1.5 Air circulation around high and low pressure areas

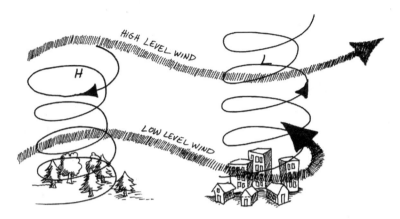

Fig. 1.6 High and low level wind circulation

At any given moment, different areas of the Earth's surface experience differences in temperature and pressure. The areas of high pressure tend to push air toward those of low pressure, causing winds, while the Coriolis Effect determines the direction of the air movement around these areas of high and low pressure. In the Northern Hemisphere, for instance, the circulation of air is clockwise around a high and counterclockwise around a low (Fig. 1.5). C.H.D. Buys-Ballot, in 1857, defined the relation between wind direction and pressure distribution, sometimes called the **law of storms**. Simply stated, in the Northern Hemisphere, if you stand with your back to the wind, lower pressure will be to your left, while in the Southern Hemisphere it will be to your right. A three-dimensional image of air circulation in the Northern Hemisphere reveals a clockwise descending corkscrew movement of air around a high-pressure area and a counterclockwise rising movement around a low-pressure area.

Near the ground, the free movement of wind is slowed by the braking force of obstacles such as hills, trees, buildings, and even the rough surface of the ground, while in the upper air regions, wind flows freely and therefore faster. This difference allows the faster upper-level winds to snake their way around high- and low-pressure areas while the slower low-level winds are drawn into low-pressure areas (Fig. 1.6).

If this picture were not already complicated enough, in the upper atmosphere, bands of very high-speed winds are generated by the large temperature differences that exist between, for instance, arctic and temperate air. The temperature difference causes pressure to push the temperate air in the arctic direction. Turned by the Coriolis Effect, these **jet streams** snake their way from west to east and, like a whipping water hose, constantly change position as the boundary between hot and cold air shifts (Fig. 1.7). With this intricate combination of air movement, temperatures, and pressure changes, it is not surprising that forecasters sometimes fail to accurately predict tomorrow's weather.

Early sailors did not have such a complete picture of global air movements. However, they had observed the general nature of convective air movements, which gave them the courage to set out on voyages that would lead them toward distant, uncharted places. The northeasterly trade winds, in 1492, blew Christopher Columbus to the shores of Hispaniola and the New World. The westerlies, in 1895 dogged Joshua Slocum as he sought to enter the Pacific Ocean from the Straits of Magellan in his solo voyage around the world. But perhaps the greatest sea voyage ever undertaken was that started in 1519 by the Portuguese navigator Ferdinand Magellan.

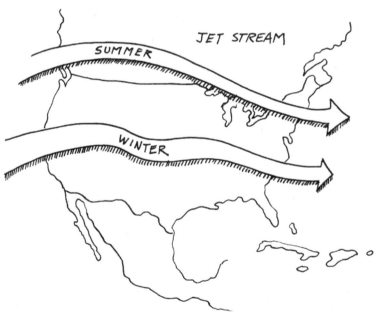

Fig. 1.7 Path of the jet stream

2

The Currents Of Patagonia
Magellan's 38-day Adventure

Where in this small-talking world can I find
A longitude with no platitude?

The Lady's Not For Burning, Christopher Fry

Mariners have always survived by understanding weather patterns and the movement of the major ocean currents without actually knowing what caused the weather and what drove the currents. They relied on natural signs to forecast weather before the availability of instruments such as barometers to measure pressure, thermometers to measure temperature, hygrometers to measure relative humidity, and anemometers to measure wind speed.

Their understanding was often summarized in proverbs that had a basis in fact. "Red sky in the morning, sailors warning; red sky at night, sailors delight" is based on the scattering of light rays in the early morning by thin cirrus clouds that, moving ahead of a warm front, often bring bad weather. Observing the smoke rise straight up from a chimney proved to be an indication of good weather because there is no wind and the temperature gradient (drop in temperature with height) is greater during good weather so the smoke is propelled upward. In bad weather, temperature changes slowly with altitude and consequently the smoke from the chimney tends to move horizontally or even to drop. In times of high humidity, sailors noticed that the hemp ropes used to hold ships' masts would shrink, causing the masts to be pulled tightly, and that the stars overhead appeared to twinkle. A drop in barometric pressure was observed by increased fatigue and joint pain, particularly for sufferers of arthritis, and by a decrease in the numbers of fish biting on a line.

Mariners also paid constant attention to changes in wind direction. They observed that a clockwise movement or change, a **veering wind**, was an indication of good weather, and a counterclockwise movement or change, a **backing wind**, indicated a nearby storm. The principal trade winds were also well known, and in some cases were the subjects of deep superstitions. In the early fifteenth century, the Portuguese Prince Henry the Navigator had a difficult time convincing sailors to go south beyond the western bulge of Africa. Most sailors at that time believed that the prevailing northerlies would prevent them from ever getting back (and that they would run into boiling waters at the equator). This problem was soon overcome with the development of the **caravel,** a new type of ship capable of beating to windward (Fig. 2.1). Before the end of the century, Christopher Columbus would sail three of these relatively small ships (the Niña was only 20 m long) to the New World, initiating decades of discoveries by other great explorers. Surprisingly, eighteen hundred years earlier, a Greek merchant named Pytheas, had sailed north from Massalia to explore England, Iceland, and Greenland in a ship almost 50 m long—gigantic by comparison to the diminutive caravels.

A final vital bit of information needed to guide sailors more quickly to their destination was the movement of major ocean currents, a fact that, as we shall see, became abundantly clear to the early European explorers seeking passage to the New World and beyond.

Fig. 2.1 A Caravel

Lucky Columbus

The Portuguese kings in the early fifteenth century were obsessed with finding an ocean route to Asia. They had heard of Herodotus's story of Phoenician sailors who circumnavigated Africa in 600 BC, sailing south from the Red Sea and returning through the Pillars of Hercules three years later. But, the map that was available at the time, which had been drawn by Ptolemy in the second century, showed Jerusalem at the center of the world with the landmass of Africa extending southward to the end of the Earth in Antarctica, uninterrupted by a sea passage (Fig. 2.2). Also, most sailors at the time were frightened to go beyond Cape Bojado in Morocco, believing that all that extended beyond the cape was the Mar Tenebroso, the Ocean of Darkness. Finally, the ships of the time, the Barchas, were sturdy, square-rigged single-masted vessels that, unfortunately, could sail only with the wind. These ships, departing from Portugal, would enter the southward flowing sea currents and be driven by the prevailing northeast trade winds, making the outbound voyage easy. The return voyage, however, required the ships to go out into the Atlantic Ocean to pick up the westerlies and the

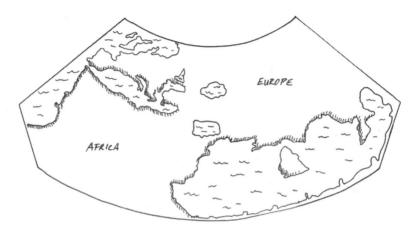

Fig. 2.2 Cosmographia of Claudius Ptolemy

eastward running current. This was a dangerous and risky proce-
dure, placing the ships out of sight of land for days with no means of
accurately determining their positions.

The development of the caravel solved this problem. Prince
Henry the Navigator had established a school in Sagres, a spit of land
jutting out into the Atlantic in southwest Portugal. In the 1440s,
shipwrights from his school came up with the idea for a ship that
could beat against the wind. It was a round-bottomed boat with a
high bow and higher quarterdeck at the stern, a shape that allowed
it to ride high in the water and not be washed by deep ocean waves.
With two, three, or four masts, the ships carried lateen-rigged triangular sails. This sail arrangement allowed them to sail close to
the wind, and provided agility and maneuverability, as well as up-
wind capability. Later versions of the caravel carried square rigging
in the fore and main masts and a lateen sail on the mizzenmast,
further improving their speed (Fig. 2.3).

Sailing these new caravels, Bartolomeu Dias de Novais succeeded in rounding Africa in 1487. He used the trick of turning southwest beyond the Congo River's mouth, sailing in a wide arc that
allowed him to pick up the westerlies that propelled him around
the continent. With this daring feat, the Portuguese had opened a
trade route to Asia.

At about the same time Cristoforo Columbo (Columbus) was
trying to convince the Portuguese court that he had found a short

route to the West whereby he could reach the Isle of Cypango, as Marco Polo called Japan. Unfortunately, since Dias had already discovered the route around Africa to the Orient and most experts were suspicious of Columbus's navigational arithmetic, the Portuguese were not interested in funding such a risky venture. There was no question of the shape of the earth, since all sailors had seen masts dip down in the horizon and were convinced that the Earth was a sphere, but the question was one of size.

It is no wonder that questions were raised about Columbus's calculations, since he had predicted that the distance from the Canary Islands to Japan was less than one quarter of its actual distance. He arrived at this mistaken conclusion in the following manner.

Eratosthenes, in about 230 BC, had figured that a degree of longitude at the Tropic of Cancer was 59.5 nautical miles. (He was less than one percent off the true distance.) Not satisfied with this result, Columbus preferred to rely on the ninth-century Moslem geographer Al Farghani's calculation. However, he mistakenly assumed that the number referred to by Al Farghani was a Roman mile, although it was actually an Arabic mile, and therefore arrived at a degree of longitude being 45 nautical miles. Columbus thus concluded that the world was 25% smaller than Eratosthenes' calculation and 10% smaller than Ptolemy's. Ptolemy had said that the known world

Fig. 2.3 Columbus's flagship, the Santa Maria

east of Cape St Vincent covered 180° of longitude while Marinus of Tyre stretched this to 225°. For good measure, Columbus then added another 28° to take into account the discoveries of Marco Polo and 30° to account for the distance from Eastern China to Japan. He therefore concluded that by going in a westerly direction, he had to cross only 68° of longitude to reach Cipangu (Japan). To further compound his error, he corrected this downward to 60° to account for Marinus's oversize degree which, at latitude 28°, he thought was only 64 km (40 miles). With these multiple errors, Columbus mistakenly thought his voyage to Japan would cover 3840 km (2,400 miles) rather than the actual 16960 km (10,600 miles). Luckily, he ran into the previously undiscovered land mass of America that was located just about where he expected to reach Japan. Yet, to convince his patrons that he had succeeded in reaching his original goal, he named the natives "Indians," and their sacred chiles "red peppers."

Columbus was lucky in another regard. On his first voyage to the New World, he arrived in October 1492, during the hurricane season. It turns out that the track Columbus took is free of hurricanes with only one even coming close in 104 years. Even in cruising through the Caribbean Islands, the odds of his encountering a hurricane were extremely low. Had he taken another track across the Atlantic or had he cruised through a different part of the Caribbean, he would most likely have encountered a hurricane or at least a tropical storm. He was lucky!

The Great Navigator

Fernão de Magalhães (Magellan) was born to a noble family in Portugal a dozen years before Columbus's first voyage to the New World. By 1511, he captained a caravel that sailed east from Portugal to the Spice Islands of the Mulaku and Banda seas, in what is now part of Indonesia. During that trip he dreamed of finding a westerly passage through Spanish South America to these same islands, a trip that he mistakenly thought would be shorter, since he had no idea how wide was the Pacific Ocean. Before he had a chance to work out the details of such a voyage, the Royal Court accused him of graft. While suffering the displeasure of his monarch, Magellan met Rui Faleiro, an expert celestial navigator. Faleiro helped him work out the theory that a line drawn around the globe (presumably a great circle), conforming to the division of the world by Pope

Alexander VI into Spanish and Portuguese areas, would pass west of the Spice Islands, placing them in a region of Spanish influence (Fig. 2.4)[4]. Relying on Faleiro's concept, and since he was already in trouble with the Portuguese crown, Magellan decided to renounce his Portuguese citizenship and approach the Spanish King, Charles V, to sponsor a voyage of exploration aimed at finding a westerly passage to the Spanish Spice Islands. The king was more than happy to do so, seeing an opportunity to expand his kingdom, and so provided Magellan with a fleet of five ships with a total of 250 men, including many officers of doubtful loyalty to the captain general of the expedition, as Magellan was called. The flagship *Trinidad*, with a displacement of 100 tons, was about 26 m long, a Lilliputian dimension compared to today's cruise ships and supertankers that are over 300 m long.

---

4   Alexander VI, an Aragonese, in 1493 gave Portugal all the land west of the 38th longitude in the mid-Atlantic ocean. The treaty of Tordesillas in 1494 shifted the line west to 46°–30', that included part of the then unexplored South American Continent.

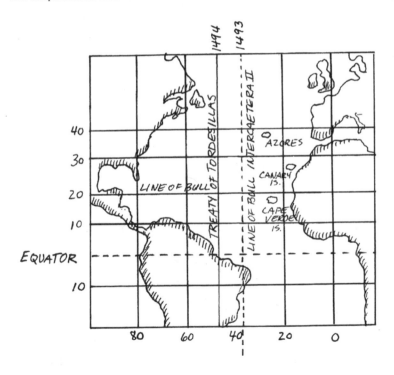

Fig. 2.4 Demarcation between Spanish and Portuguese domains

On Tuesday, September 20, 1519, Magellan set sail from Sanlú-car de Barrameda on the Spanish coast, on a course to the Canary Islands. But the voyage started on a bad omen: Faleiro decided not to go along, believing his horoscope, which forecast a fatal jour-ney. After a brief stopover in the Canary Islands, the fleet continued southward, skirting the African coast before turning toward Brazil. Unfortunately, they had followed the coastline too far south and entered the **equatorial doldrums** that becalmed them for many days, furnishing an opportunity for a mutiny, plotted by one of the disloyal captains, a mutiny that was firmly put down by Magellan.

Currents, driven by the spinning of the Earth and the constant winds that blow north and south of the equator, are like invisible rivers steadily moving in the oceans (Fig. 2.5).

Magellan's little fleet was caught in such a current moving west-ward toward the coast of Brazil that guided his ships toward the southeast trade winds. On December 13, the fleet entered the har-bor of Rio de Janeiro, having bypassed potentially unfriendly Por-tuguese settlements farther north. After subduing a second mutiny and re-provisioning his ships, Magellan continued down the coast looking for the strait leading to the Indies that he had seen on a chart reputedly drawn by Martin of Bohemia, a pilot and cosmogra-pher. After a brief stop in Montevideo, the fleet continued along the coast, helped by winds from the northwest that blew in that season, and finally put into Puerto San Julián, where the fleet wintered until August 24th. Magellan called this land *Patagonia* after meeting a big-footed (patagón) native.

Before they had a chance to settle in for the winter, the expedi-tion almost fell apart as three of the captains mutinied and took control of their ships with the intent of forcing Magellan to return to Spain. With the support of loyal ordinary seamen, Magellan firmly took back control of the ships and court-martialed the three cap-tains, condemning two to death. The third captain, who had been spared the death penalty, was soon caught trying again to stir up the men and foment another mutiny with the aid of a chaplain. This time, Magellan had enough and marooned the captain and chaplain before continuing on his voyage. However, he now had one less ship, as the *Santiago* grounded and could not be re-floated.

Magellan's chronicler, Antonio Pigafetta, describes the discov-ery of the strait. Passing the Cape of Eleven Thousand Virgins on

Fig. 2.5 Major ocean currents

October 21, "...the Captain General found it. He knew where to sail to find a well-hidden strait, which he saw depicted on a map...." But was it really a strait or simply a long dead-end bay? The weather contrived to provide the answer as a northeast gale drove two of the remaining ships through the first narrows into a large bay from which they saw fires burning on the shore at night, and so called the land *Tierra del Fuego*. As the weather abated, the two ships sailed on through the next narrows to a broad bay, which led them to believe that they were indeed in the strait and so, with great joy, they sailed back to announce their discovery to Magellan. For thirty-eight days Magellan and his ships cautiously explored the twists, turns, and dead ends in the 334-nautical-mile-long strait, often fighting strong currents, but luckily encountering none of the powerful squalls, called williwaws, that other sailors have endured in these waters (Fig. 2.6). Joshua Slocum, the first man to sail solo around the world described them as, "...compressed gales of wind that Boreas handed down over the hills in chunks. A full-blown williwaw will throw a ship, even without sail set, over on her beam ends." Although Magellan had an easy passage through the strait, he was betrayed during this time by one of his captains, who finally led a successful mutiny, turning one of the remaining ships back toward Spain. Magellan was thus left with only three ships that on November 28, passed out of the strait to embrace the broad, open, Pacific Ocean, so named because it appeared to Magellan to be a calm sea with fair weather compared to the familiar, rough, Atlantic Ocean.

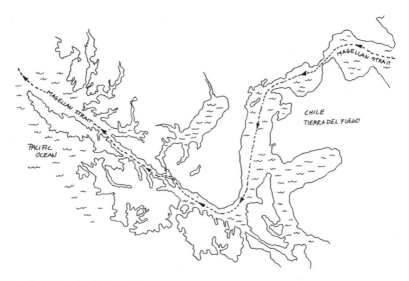

Fig. 2.6 Strait of Magellan

But, Magellan had not a clue to the immensity of the Pacific Ocean since he mistakenly believed, as did Columbus, that the globe was much smaller than it is actually and furthermore, he had no accurate way to determine longitude, the angular distance around the globe from a meridian drawn through Greenwich, England (Fig. 2.7). Accurate longitude determination had to await the establishment of the Greenwich observatory in 1675, its observations of the moon starting seventy-five years later and, in the eighteenth century, the invention by John Harrison of the chronometer, an instrument to accurately measure time at sea. However, in the sixteenth century, sailors could determine latitude, which is the angular distance north or south of the equator, by using an astrolabe, a crude instrument that measures the angle between the horizon and any heavenly body, such as a star or planet. Its first use is believed to date back to Hipparchus in the second century BC. Since Polaris, the North Star, is relatively stationary in its position over the North Pole, latitude could be determined by using that star as a reference (Fig. 2.8). Of course, a small correction had to be applied to the observation to account for the fact that Polaris is not exactly over the North Pole but actually circles around it. During daytime, latitude is determined by measuring the altitude of the sun at noon, since its elevation at defined locations and

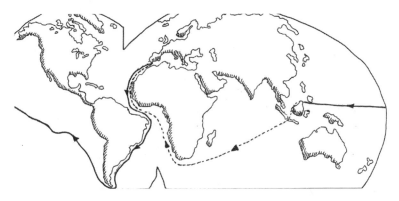

Fig. 2.7 Magellan's route

the changes throughout the year are recorded for navigators in an ephemeris. This table, giving the daily location of selected celestial bodies, was available even in the sixteenth century. In the Southern Hemisphere, which was not as familiar to European explorers, the Southern Cross points to a faint star lying almost directly over the South Pole and serves as the heavenly reference point for determining latitude at night (Fig. 2.9).

99 Days Later

Leaving the strait, Magellan's three ships followed the northwestward prevailing winds and the northward flowing Humboldt Current that combined to keep them close to the coast of Chile for almost three weeks before they picked up the southeast trade winds that would carry them across the Pacific. While the expedition was incredibly lucky with the weather, experiencing fair winds, deep blue seas, and only the hint of puffy, white, fair-weather clouds, the unexpected duration of the voyage created an extreme hardship; Magellan still believed that the trip to the Spice Islands should take no more than a few days. Instead it took about four months, during which time biscuits had turned to powder and were crawling with worms, the remaining water was putrid, and the crew took to eating rats and chewing on rawhide, all the while suffering from bleeding gums and general anemia, the ill effects of scurvy (from a lack of Vitamin C). Eleven men died on the voyage before the first island was spotted.

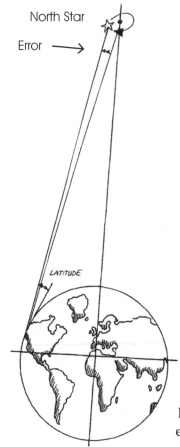

North Star

Error $\longrightarrow$

LATITUDE

Fig. 2.8 Path of North
Star and the error

Almost two months after the ships left the strait, Puka Puka, an uninhabited, wooded but uninviting postage stamp of an island, proved unapproachable because it had no safe anchorage. At this time, Magellan may have believed that he was near his goal, the Spice Islands, and continued on his way. Except for a few small islets, where he stopped to catch sharks, the Pacific offered no sanctuary for another two months until the green hills of Guam came into view on March 6th and the weary crew sailed into Umatak harbor. They were immediately surrounded by Chamorros (Polynesians who had conquered the island centuries earlier), who raced around in their outrigger canoes stealing anything they could reach, even skiffs hanging from the stern of one of the ships. After having subdued the natives, Magellan stayed in Guam only long enough to re-provision his ships.

The little fleet continued westward, arriving in the Philippines on March 13th, thus giving Magellan the title of the first man to have circumnavigated the globe, since on an earlier trip he had sailed eastward as far as Ambon in the Indonesian islands (in the Banda sea). However, Magellan was not to return to Spain. Having had success in converting one local tribe to Christianity, Magellan was refused by another. In retribution, he burned their village and ordered them to supply his ships with three goats, three pigs, three loads of rice, and three loads of millet. They refused and offered

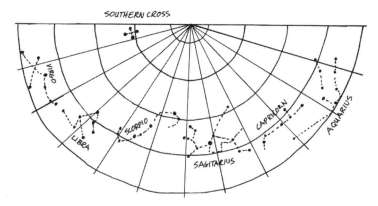

Fig. 2.9 Star Chart of the Southern Hemisphere

instead to send two of each. This compromise did not satisfy Magellan who, on April 27, 1521, with sixty of his men, attacked the tribe on the island of Mactan. After a fierce battle, Magellan and six of his men were killed.

Of the original fleet, only two ships left the Philippines, the third having been abandoned as too worm-eaten, and in any event the 115 survivors were simply too few to crew three ships. Stopping to load a cargo of spices in the Moluccas, the two remaining ships headed in different directions, the *Trinidad* heading for the Pacific Ocean and the other continuing westward. The *Trinidad* was captured by a Portuguese fleet and ended up destroyed when struck by a squall while at anchor. Of the five ships that left Spain thirty-seven months earlier, only one, appropriately named the *Victoria*, returned, with a crew of only eighteen but with a ton of valuable spices on board. The captain, Juan Sebastián de Elcano, was heralded as the first to circumnavigate the globe, and was generously rewarded for his accomplishment by Charles V. The rewards included the grant of a coat of arms embellished with two cinnamon sticks, three nutmegs, and twelve cloves. But it was Magellan who, two centuries later, would be credited by historians as the man with the vision and the fortitude to undertake one of the greatest voyages of discovery.

The Eunuch Mariner

Neither Columbus nor Magellan would have been able to navigate the open seas were it not for a Chinese invention that allowed them to steer a given course even in fog or at night: the magnetic compass. Crude compasses first appeared in China in the fourth century bc but the first seagoing compass dates from the twelfth century. China, in the early first millennium, was a major maritime power in Southeast Asia and employed great shipbuilders. In 1404, Admiral Zheng He (1371–1435), a Muslim eunuch employed by a prince of the Ming dynasty, assembled a fleet with as many as 300 ships that included a giant nine-masted vessel, 142 m long by 58 m wide (475 ft × 193 ft), six times larger than the miniscule caravels that carried Columbus to the New World. With this mighty fleet, Admiral Zheng undertook voyages for twenty-eight years, from China to India to the east coast of Africa and, it is believed, he even rounded the Cape of Good Hope and reached as far north as Europe—possibly even as far west as the Caribbean. His voyages demonstrated the power of the Chinese civilization and led to the formation of many ties with other nations. Unfortunately, after Zheng's death, most records of his voyages were destroyed and a ban was instituted by the Emperor on the construction of vessels with more than three masts, effectively strangling the future development of China as a major maritime power for five hundred years. But Zheng's legacy survived because of the numerous innovations in shipbuilding that he originated, including central rudders and watertight compartments. More important, perhaps, were 24 maps drawn by Sambao, who sailed with Zheng. The maps chronicled the voyages and accurately depicted the world they traveled.

3

Force Ten
The Sunken Treasures of the Buccaneers

> Blow, winds, and crack your cheeks! Rage!
>   Blow!
> You cataracts and hurricanoes, spout
> Till you have drench'd our steeples,
>   drown'd the cocks!
>
> *King Lear*, Shakespeare

The ocean is not always as kind as it was to Magellan on his epic voyage. Driven by the force of the wind, the seas can swell and rage, washing over the decks of ships running into the waves, or can even capsize a vessel that is turned broadside to a wave. The bottoms of the oceans are littered with ships that have been bested by such angry seas. In the seventeenth and eighteenth centuries, buccaneers sailed throughout the Caribbean and the eastern shores of the Americas, attacking ships laden with treasures destined for European royal coffers. Both the treasure-rich galleons and the buccaneers' ships were often sunk during storms by either being driven into reefs or overwhelmed by high seas. Treasure hunters seeking such lost riches have eagerly explored these ships. But, before we follow the ill-fated voyage of the Spanish galleon *Atocha*, we must understand how a storm is born.

Out of Mythology

With the body of a bird and the face of a woman, the Greek goddesses of storms, the *Harpies*, were both feared for the damage they could inflict and respected for their contribution to the fertility of plants and animals. The Greeks also assigned a wind god to each of the eight points of the compass. The northern wind, Boreas, was the strongest and most violent, causing storms and heavy seas, but having destroyed the enemy fleets of Xerxes and the Persians, was also revered as a protector of the city of Athens.

Each ancient culture worshiped wind gods characterized by strength and violence, simultaneously the healer and the demon, the nymph with the body of a serpent, the strongman with wings of an angel. In the north (Iceland and Norway), the Teutons worshiped Thor, the powerful god of thunder and storms. Behind all these symbolic representations there was the realization that the wind was both the giver of life that brought rain to nourish the plants and the destroyer, whose tempests could uproot trees and sink ships.

The scientific investigation of wind began in Greece. Aristotle wrote his treatise, *Meteorologica*, late in life. He was hampered in his investigations by a lack of data that could be obtained only from instruments to measure temperature, wind velocity, and atmospheric pressure. Most of these would not be invented for almost two thousand years; the thermometer by Galileo in 1607, the barometer by Toricelli in 1643. Those and the four other instruments directly connected with the understanding of weather (telescope, microscope, air pump, and pendulum clock) were all invented during the lifetime of René Descartes (1596–1650), the scientist who recognized that water droplets coalesced and, when they became too heavy, fell as rain or snow.

How Fast the Wind Blows

Until the development of instruments to measure wind speed, the intensity of the wind was described by seamen in a qualitative manner. In fact, such terms are commonly used even today to describe the strength of the wind. A mirror-like appearance characterizes calm seas, while a gentle breeze causes large wavelets with some foamy crests. In a strong gale the waves are high and there are streaks of foam and spray hindering visibility. In 1805, Sir Francis Beaufort, an admiral of the British Royal Navy, proposed a scale from 0 to 12 that defines the full range of wind speeds from calm to hurricane force, and is now know as the Beaufort Scale. To define super typhoons, such as the 185 km/hr Typhoon Manchu that struck the region of the South China Sea, the scale was extended to 17.

Beaufort Number	Wind Speed		Description
	m/s	mph	
0	0–0.2	0–1	Calm
1	0.3–1.5	1–3	Light air
2	1.6–3.3	4–7	Light breeze
3	3.4–5.4	8–12	Gentle breeze
4	5.5–7.9	13–18	Moderate breeze
5	8.0–10.7	19–24	Fresh breeze
6	10.8–13.8	25–31	Strong breeze
7	13.9–17.1	32–38	Moderate gale
8	17.2–20.7	39–46	Fresh gale
9	20.8–24.4	47–54	Strong gale
10	24.5–28.4	55–63	Whole gale
11	28.5–32.6	64–72	Storm
12	32.7+	73 +	Hurricane

In order to grasp the nature of storms, it is first necessary to realize how water creates clouds. The air around us is never perfectly dry—it always contains some humidity in the form of water molecules. When a given volume of air cannot hold any more water molecules, we say that it is saturated. As temperature increases, the amount of water needed to saturate the air also increases. The **dew point** temperature is the point below which air that has a given amount of water vapor begins to condense it out as a liquid (Fig. 3.1). If, for instance, the air is almost fully saturated when the outside temperature is 30°C and the temperature drops overnight by

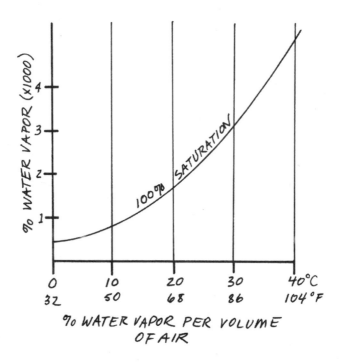

Fig. 3.1 Water vapor

5°C, then, in the morning, we will find the ground covered with dew because at 25°C the air was 100% saturated.

Birth of a Storm

A **storm** is born as warm humid air, which is lighter than dry air above, begins to rise from the ground in an **updraft.** As it rises, this air packet expands and consequently cools. When it reaches its dew point temperature, condensation takes place, forming a cloud. This condensation releases heat, which causes the temperature-drop to decrease with height. Driven by the increase in buoyancy, the cloud is pulled ever higher and is joined by more packets of rising humid air, enlarging the cloud. At very high altitudes, water droplets[5] or ice crystals within the cloud may grow big enough to overcome the force of the updraft and begin falling. As these descend, they drag along the air, causing **downdrafts** and precipitation. As rain be-

---

5   As they freeze, water droplets release more heat.

gins to fall, evaporation takes place, causing the downdraft to cool and the rain drops to further accelerate. In its most violent configuration, the storm may have a combination of updrafts and downdrafts, together with thunder and lightning. Such thunderstorms can rise to a height of about 13 000 m (43,000 ft) but rarely extend horizontally more than 24 km (15 miles).

But from such modest beginnings, a **tropical depression**—a storm that forms in the tropical region—can grow into a monster storm spiraling over an area of 1,280 km (800 miles). When such a storm forms in the Atlantic Ocean, it is called a **hurricane,** from the Carib word **urican,** for big wind, and if in the Pacific, a **typhoon,** from the Chinese **taifeng** (Fig. 3.2). In the Indian Ocean and the Coral Sea (between Australia and New Guinea), these storms are called tropical **cyclones** which may be the most accurate terminology based on their appearance as viewed from space.

Occurring in the summer or fall, a hurricane is born as a **tropical depression,** a region of low pressure that spawns thunderstorms driven by warm, humid air rising from the ocean. It may originate off the coast of West Africa or anywhere between there and the Caribbean. At first, a line of thunderstorms starts to spiral in a counterclockwise direction toward what will become the hurricane's eye, a center that is washed clear of clouds by sinking air. As the storms intensify and the cyclonic flow increases, wind velocities reach 33 mps (74 mph) and the area of the disturbance radiates out

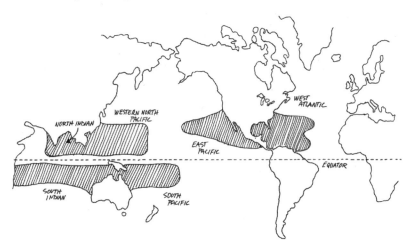

Fig. 3.2 Where are the storms?

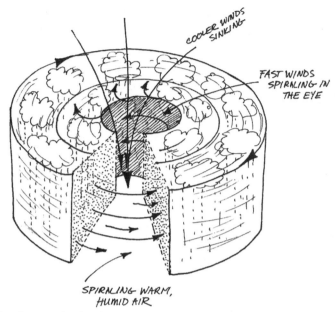

COOLER WINDS SINKING

FAST WINDS
SPIRALING IN
THE EYE

SPIRALING WARM,
HUMID AIR

Fig. 3.3 Anatomy of a hurricane

as much as 800 km (500 miles) (Fig. 3.3). In the Atlantic Ocean, a hurricane travels east to west and is driven around an area of high pressure that lingers in the region of the island of Bermuda during the months from August to October, and is affectionately called the Bermuda High (Fig. 3.4). Once a hurricane moves far enough west, the clockwise circulation around the Bermuda High carries it northward and eastward. As a hurricane passes overhead, waves of increasing wind and rain are felt as the bands of thunderstorms move across the area with mounting intensity. The most severe upward spiraling winds are along the edge of the eye.

Apart from the heavy rain and high winds, a hurricane may also cause a **storm surge** with giant waves that crash on shore, washing away beachfront and even riverfront homes. To understand how such a storm surge originates, imagine the ocean's surface under a hurricane. Air pressure decreases when moving from the outer edges toward the eye. The ocean's surface responds to this variation in pressure with the outer edges being pushed down and the area of the eye being sucked up. It is as if the moon's gravitational pull on the oceans were focused on a small area. The mound of water under the eye may be only one meter (3 feet) high in mid ocean, but as the eye of the hurricane approaches land, the mound under it becomes higher because the rising ocean floor prevents water from flowing out. If a surge occurs at high tide, the combined effect is disastrous. At Pass Christian, Miss., Hurricane Camille came ashore in 1969 with a 7.2 m- (24 ft)-high storm surge, destroying thousands of homes and forever altering the shoreline—at least until the arrival of the next hurricane.

Caribbean Tragedy

The failed conquest of England by the great Spanish Armada of 1588 signaled the start of a gradual political and economic decay of Spain that stood in sharp contrast to the golden age of its science, art and literature. At the beginning of the seventeenth century, Cervantes had published the immensely popular *Don Quixote,* about the transforming power of illusion, and for a short period, Spain was insulated from European wars. During this time, both England and Spain exploited the New World for treasures to finance their political escapades. Sir Walter Raleigh, to appease his queen (after an affair with one of her ladies in waiting that resulted in his imprisonment and banishment from the court), sailed in

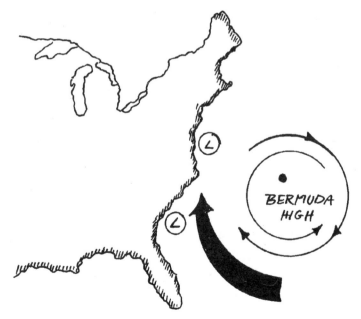

Fig. 3.4 Path around a Bermuda High

search of an imagined gold mine in Guinea with a fleet of ten ships with 90 gentlemen and 318 cutthroats. Failing in his seven-month-long mission, Raleigh was then tried for treason and executed on his return to England. James I handed Raleigh's head on a charger to the Spanish ambassador, Diego Sarmiento de Acuña Gondomar, who presented it to his king, Philip III, as a gift from the English monarch. It was considered retribution for the many insults Raleigh, during his life, had inflicted by attacking the Spanish fleets. In that same year, 1618, Philip embroiled his empire in the Thirty Years War, defending fellow Hapsburgs from Protestant incursions in the German states and to halt Dutch expansion in the Caribbean and Indian Oceans. By the time his son, Philip IV, ascended to the throne in 1621, Spain was desperately in debt. It was, therefore, vital for Spanish silver from the Peruvian mines to flow continuously to fill the shrinking imperial coffers. The silver mines of Potosi, on a high Peruvian plateau, had been in operation since Francisco Pizarro conquered the region over seventy-five years earlier, and provided silver that was shipped with the annual fleets that plied the route between Spain and the Caribbean. These silver fleets sailed twice a year, bringing the plundered riches of gold and silver from the New World to fill the coffers of their European masters.

To transport the Peruvian silver, large fleets were assembled that were heavily armed to guard against attack by Spain's enemies as well as by marauding pirates. After all, a million-peso cargo was well worth protecting. The fleet that assembled in Portobello in the summer of 1622 included the 500-ton galleon *Atocha* under the command of Captain Bartolomé Garcia de Nodal, and a new ship, the *Santa Margarita*. All together, 28 ships assembled in Portobello, on the northern shore of Panama. They were met by mule trains carrying treasure including both gold and silver ingots and boxes of jewels that had been brought across the isthmus from the Pacific shore, where it had been shipped from Peru.

These caravans of up to one hundred mules trekked along trails guarded by armed men, many of whom died of fever in the infested jungle. It was along this trail that Francis Drake successfully carried out an ambush of a Spanish mule train. Drake sailed his fleet under the protection of Elizabeth I and would later undertake one of the great sea voyages, taking the *Golden Hind* around the world, the first English ship to do so.

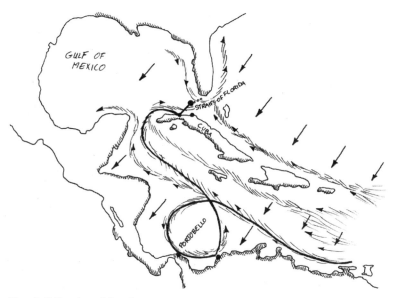

Fig. 3.5 Route of the Spanish fleet of 1622

The Spanish fleet, with its treasure safely in the holds of the ships, left on July 22 and, following the sea current, stopped in Cartagena to take on more treasure (Fig. 3.5). The annual fleets that served the Spanish Main usually followed a routine of first calling on Margarita Island to take on pearls, then continuing to Portobello (Nombre de Dios bay) where they picked up the bulk of their treasure. They then continued to Havana, their last port of call in the Americas, occasionally stopping in Santo Domingo, which was the administrative capital of the region.

In that summer of 1622, the fleet went directly to Havana from Portobello, arriving there on August 22. There, they delayed their departure for a week, having been convinced by the pilot that the coming conjunction of the earth, sun and moon on September 5 was a sign of approaching severe weather. Nevertheless, they departed on September 4, a gloriously clear day, but still dangerously in the Caribbean hurricane season.

While they had waited in port, a weak circulation formed northeast of the Leeward Islands and slowly intensified into a tropical storm.

The ships sailed north toward the Florida Keys, where they expected to encounter the **Gulf Stream** that would carry them home.

On the first day at sea, cumulus clouds were seen in the southeast and the setting sun shone bright red, a sign that every sailor knew meant bad weather. The wind intensified during the second day, raising heavy seas that tossed the proud fleet about, causing some of the ships to lose their rudders and others to lose their masts, even though most sails had been stowed. The sky became increasingly black and many of the ships in the fleet lost sight of each other. One by one, the ships were propelled by the furious winds toward the reefs that guard the Florida Keys. While the passengers and crew of the *Atocha* prayed for their eternal souls, the ship was carried by a wave high above a reef and dropped violently onto it, breaking open her hull, causing sea water to rush in, and drowning all who were trapped below deck as she sank to the bottom of the sea. Of 265 souls aboard, only five survived, and out of the fleet, six ships, including the *Margarita*, were lost, together with their bounty and the hopes of the Spanish court.

Since she had sunk in relatively shallow waters, a salvage effort was immediately mounted to retrieve the lost treasure. Unfortunately, a second hurricane struck as these efforts were under way and tore the *Atocha*'s hull apart, scattering wreckage over a large area and burying the precious treasure under a deep blanket of sand.

Three hundred and fifty years later, after a sixteen-year search, the treasure was recovered west of the Marquesas in a modern-day salvage operation, and proved to be the richest collection of lost treasure ever found.

4

Hurricane Strength
No Region is Immune

> Yes, one of the brightest gems
> in the New England weather
> is the dazzling uncertainty of it.
>
> *About Weather in Hartford*, Mark Twain

L ike the *Atocha,* ships caught at sea in the path of a hurricane must fight a terrible onslaught of wind and rain but may have a chance to survive while "rolling with the punches." However, once over land, a strong hurricane destroys almost everything in its path: fragile buildings that cannot move out of the way; trees that can bend only so far. Even some vehicles may be picked up by the wind to be tossed about like leaves, and people caught out in the open, especially near the water's edge, are in danger of drowning in the storm's surge.

One of my earliest memories after arriving in the United States is of driving through New England a year after the 1938 hurricane.

There were downed trees everywhere, centuries-old growth, wiped out in minutes. Power and telephone cables lay across the landscape, having been braided by an awesome force. As I stared out of the car window, I wondered how the wind could wreak such damage.

How Strong Is It?

The first question that comes to mind when describing a storm, an earthquake, or a hurricane is: How strong is it? Most measures of natural events started out being described qualitatively by noting, for instance, what happens to objects: a chandelier swinging in an earthquake or a tree limb bending and then breaking off in a hurricane. In the early 1970s, Robert Simpson and Herbert Saffir developed a scale defining the strength of a hurricane based on measured quantities. It is known as the **Saffir-Simpson Damage Potential Scale**. Today, hurricanes are sometimes described as Beaufort scale 12–16 with Beaufort 12 equivalent to Saffir-Simpson 1 etc.

Saffir-Simpson Damage Potential Scale

Category	Wind Speed	Storm Surge	Barometric Pressure
	m/s (mph)	m (ft)	millibars (inches)
1 Minimal	33–42 (74–95)	1.2–1.5 (4–5)	>980 (>28.94)
2 Moderate	43–49 (96–110)	1.8–2.4 (6–8)	965–979 (28.50–28.91)
3 Extensive	50–58 (111–130)	2.7–3.6 (9–12)	945–964 (27.91–28.47)
4 Extreme	59–69 (131–155)	3.9–5.4 (13–18)	920–944 (27.17–27.88)
5 Catastrophic	>69 (>155)	>5.4 (>18)	<920 (<27.17)

Based on this scale, it is quite likely that the hurricane responsible for the sinking of the *Atocha* would rate as a Category 3 or less, since there is no contemporary historical record of extensive damage to the Caribbean Island settlements. Category 5 hurricanes are, fortunately, relatively rare, with only three making landfall since the turn of the twentieth century: in 1935 and 1969 and most recently, Hurricane Andrew in 1992. The frequency of major hurricanes is directly related to the temperature in the North Atlantic (as well as the phenomenon described as El Niño [Chapter 13]). During warm periods, there are more major hurricanes, as illustrated by the periods 1926–70 and 1995–2000[6]. Cool periods of 1903–25 and 1971–94 had few major hurricanes. A majority of the hurricanes that pass over the eastern United States hit the southern coast before veering out over the Atlantic, losing their strength over the cooler northern waters. However, a unique meteorological condition favors another scenario.

New England, 1938

French meteorological observers at the Bilma oasis in the Central Sahara Desert (in today's Niger) noted a wind shift on September 4th 1938. Such a change in wind direction is normally associated with an upper air depression or area of storminess that in this case was about to pass west of Dakar on Africa's West Coast toward the Cape Verde Islands in the Atlantic Ocean. In fact, bands of thunderstorms grew there and a cyclonic circulation formed between the northeast trade winds, the southwest monsoon, and the equatorial easterlies. Now classified as a tropical depression, the thunderstorms were organized into an ever-growing swirl. Day by day, the winds increased in intensity until reaching gale strength, at which time they became a tropical storm. The gathering storm strengthened as it moved westward, and on September 16th was reported to have reached hurricane strength by a freighter, the *SS Alegrete*, northeast of Puerto Rico. Three days later, the weather bureau in Jacksonville, Florida, issued a warning that an impending severe storm was expected to strike the Miami area within twenty-four hours. However, this particular storm had different plans.

We have already described how the high-pressure area known as the Bermuda High steers hurricanes around it like a sling shot. But, when a continental high-pressure area over the Continental United States moves eastward, and squeezes a low-pressure trough between it and the Bermuda High, a restricted slot is created through which a hurricane has to pass. Just as the narrowing of a river increases its velocity, a hurricane passing through such a slot is propelled faster

---

6   The Atlantic Ocean warmed by 1–3°F over the last decades of the twentieth century, setting the stage for hurricanes to become more severe in the early twenty-first century.

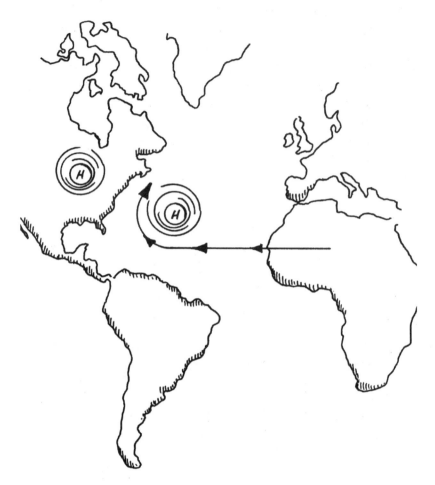

Fig. 4.1 Trajectory of the 1938 New England Hurricane

and aimed northward, directly toward the New England coast (Fig. 4.1). This rapid movement also retains the strength of a hurricane that would normally become weaker as it passes slowly over the cooler northern waters. Not since 1815 had a storm of this magnitude been aimed directly at the populous northeastern states.

Heavy rains soaked New England for several days before September 21, 1938. That morning, the hurricane that had turned northward, avoiding the Florida coast, was located about 160 km (100 mi) east of Cape Hatteras, NC. It was now a Category 3 storm but was expected to sweep eastward and pass harmlessly out to sea. However, it was caught in the low-pressure trough between the Bermuda and Continental highs, causing it to suddenly speed northward at an incredible 80 km/hr (50 mph). It reached Long Island, east of New York City, 560 km (350 mi) away, within seven hours. Low pressure in the center of the storm lifted the sea level as much as 5 m (17 ft) above high water, while the fast-moving storm pushed this surging sea toward shore in a wave that some compared to a tsunami[7]. Buildings all along the coast were demolished as the ocean swell swept well inland, smashing everything in its path with brutal swiftness. Water rose to the second floor of downtown businesses as the surge moved up the Providence River. In Norwich, 21 km (13 mi) from the sea, there was 3.6 m (12 ft) of water in the center of town, and a mother huddled with her three children in the cellar of their home thought, "it was the end of the world."

Curious observers who came to watch the spectacle along the oceanfront were engulfed and drowned by the surging sea. The fishing village of Montauk, on the eastern tip of Long Island, was virtually destroyed, and all along the southern New England coast, houses were swept away, carried by wind gusts of up to 83 mps (186 mph) as measured outside of Boston. At Watch Hill, RI, thirty-nine occupied houses were swept away with fifteen of the forty-two occupants surviving the ordeal as their houses were washed across the bay. One survivor, Helen Joy Lee, wrote "I saw our three-car garage lifted up and dropped into the bay..." and then she herself was washed into the bay. "...The side of a house came along; there were two holes where windows had been. I could not get on it...and I was about ready to give up, when a piece of house came by and I crawled onto that." Before reaching land, she was hit by flying debris, leaving her body cut and bruised, and was blasted by sand carried by wind-blown spray, and nearly drowned.

Boats all along the coast were pulled from their anchorage and over 2,600 were destroyed, either by being capsized or driven onto land. Two thirds of the fishing fleet on Block Island was wrecked and thrown up onto the stone jetty like so many match sticks. The shoreline rail connection between Boston and New York was cut in numerous places by washouts and downed bridges, and remained closed for weeks after the storm.

---

7   See, *Why the Earth Quakes*, Levy & Salvadori

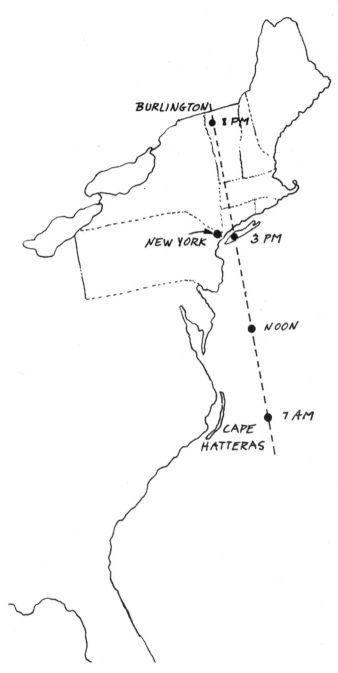

Fig. 4.2 The path of destruction

Its northward movement slowed only slightly on its path over land. The eye of the storm reached Burlington, Vt., 430 km (275 mi) away, six hours later (Fig. 4.2). On its way north, it uprooted and downed 275 million trees, smashed 26,000 cars, cut one third of all telephone lines and over 90% of power that relied on overhead lines. In Hartland, Vt., a hundred-year-old covered bridge was blown away by the wind. "The wind was so strong it mowed down them white pines just like hay," recalled John Bowley. "Most people think the flood took the bridge out, but the wind just blew it over."

Throughout the region, heavy downpours with up to 430 mm (17 in) of rain accompanied the storm and were added to the already saturated ground. Floodwaters filled the streets of Hartford and Springfield in the Connecticut valley, and every riverside town throughout New England was also flooded. Dams were overtopped and breached, releasing their stored water into rivers that were already flooded.

The morning after the passage of the storm, the residents of Cambridge, Mass. awoke to a scene of incredible devastation. Every tree had been knocked down, including the old elm tree on the commons that had stood witness to George Washington, on July 3, 1775, drawing his sword as he wheeled his horse to face the assembled troops, thus signifying his assumption of command of the continental army. On Mount Washington, the highest peak in the eastern United States, a long section of the trestle that carries the historic cog railway was blown down by winds that gusted up to 73 mps (163 mph)[8].

More than six hundred people died as a result of the storm, the greatest weather-related loss of life since the disastrous 1900 Galveston, Texas storm that claimed six thousand victims. The storm-inflicted damage resulted in losses that, measured in current dollars, would have exceeded that resulting from 1992's devastating Hurricane Andrew but are far less than that suffered as a result of 2005's Hurricane Katrina (see p. 155).

Storms of such intensity are rare, although based on the history of the past three hundred years, of the five to ten hurricanes to

---

8   As I traveled north from New York to New Hampshire in the summer of 1939 I clearly remember seeing massive trees that had been snapped through their trunks lying naked along the edges of the road. For many years thereafter I would see these fallen trees slowly reabsorbed into the returning forest. It is the same healing process undertaken by nature following devastating forest fires and volcanic eruptions.

strike New England, at least one that matches the destructive power of the 1938 hurricane can be expected every 150 years.

No region of the world is immune from storms. When suddenly forming over land, some storms will twist and turn....

5

Dorothy's Whirlwind
Great Plains Tornadoes and Cyclones

From the far north they heard a low wail
  of the wind,
and Uncle Henry and Dorothy could see
 where the long grass bowed in waves
  before the coming storm.
There now came a sharp whistling in the
  air from the south,
and as they turned their eyes that way
 they saw ripples in the grass coming from
  that direction also...
There's a cyclone coming...

The Wizard of Oz, L.Frank Baum 1900

On a late spring morning, the sun rises over the Great Plains of the United States, slowly warming the cool morning air. The sky is cloudless and somewhat hazy blue, unlike the crisp, sharp blue of a winter's morning. As the sun begins to heat the ground during the morning hours, a layer of warm air develops above the earth's surface that is reinforced by warm, moist air sweeping up from the Gulf of Mexico. Driven by convective currents, buoyant columns of invisible air begin to rise, and as they detach themselves from the earth's surface, they become bubbles of air. Within these air bubbles, thermal currents swirl upward and outward from the center as the air rises in the center and sinks along the sides. Suddenly, as the bubbles break through the condensation level, they reveal themselves as **cumulus clouds**[9] with their characteristic nearly flat bottoms and grey-white, puffy tops (Fig. 5.1). Pushed by the wind, these clouds appear to float about at altitudes that are generally below 1 800 m (6,000 ft).

As the air temperature increases toward midday, the base of the cumulus clouds rises and some of them may dwindle and disappear while others increase in size. By mid-afternoon, the remaining cumulus clouds are widely separated but individually much larger. Fed by the warm, humid Gulf air, the remaining clouds mushroom higher and higher, driven by powerful updrafts.

Some of these clouds may extend high into the troposphere and are like the thunderheads that form the seeds of the hurricanes described in Chapters 3 and 4. Called **cumulonimbus**[10] clouds because they produce rain, their tops extend above the freezing level, allowing formation of ice crystals that will eventually fall to earth

9   From the Latin word for heap
10   From the Latin, violent rain

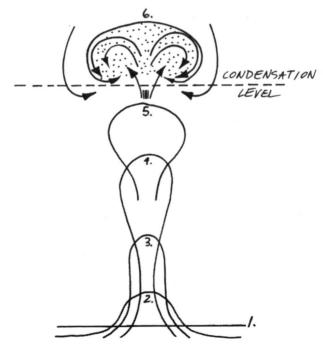

Fig. 5.1 Growth of a Cumulus

as rain. In warm tropical air, the top of a cumulonimbus cloud can reach an altitude of 15 000 m (50,000 ft), while in cold arctic air its top may be only half as high. This difference in height is partly explained by the fact that warm air can hold more moisture than can cold air, and is the reason why a tropical thunderstorm produces heavier rainfall. Storms with the strongest winds actually occur most frequently in the midwestern United States because there is dry unstable air above, coming from the desert mountain west.

Viewed from the side, a cumulonimbus appears like a churning white-to-black mass with a cauliflower surface and a top resembling an anvil (Fig. 5.2). Rain may fall from one side of the base while the other side is still sucking up air. Furthermore, lightning may be visible within the cloud and between the cloud and the ground. In the Central Plains of the United States, a unique set of circumstances comes together that may reveal an especially violent manifestation of the storm.

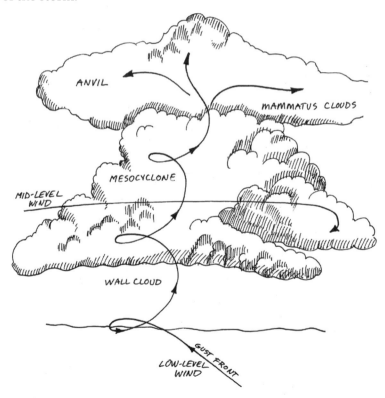

Fig. 5.2 Anvil form of the top of a cumulonimbus

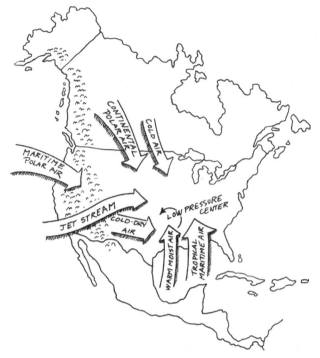

Fig. 5.3 Tornadoes formed by air masses meeting over the central United States

Tornado

A wedge of cool, dry, unstable Polar air flows down from the plateaus of the Rocky Mountains in the West and spreads over the warm tropical air flowing across the Plains from the Gulf of Mexico (Fig. 5.3). This disturbance to the smooth warm air flow is further stirred up by the **jet stream**, a fast-moving river of high-level air arriving from the west-southwest, and results in a turbulent atmosphere (Fig. 1.7). The combination of these two events—instability and shear—results in a rapid cyclonic circulation developing within the cumulonimbus. This causes the in-rushing air to spin faster and faster, dropping the pressure within the vortex. In turn, this produces a cone of condensation that hangs below the cloud base. A **tornado** develops from this spinning funnel, with a rapidly spiraling updraft on its surface and a weaker downdraft in its interior (Fig. 5.4). The exact mechanism that initiates a tornado is not yet

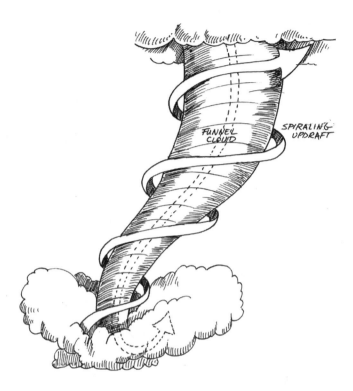

Fig. 5.4 Structure of a tornado

known, although scientists suspect that a downdraft from the base of the cloud swoops out along the ground toward the low-pressure zone beneath the vortex and is swept up. The current theory for tornadoes in "supercell" thunderstorms is that the air that eventually feeds the tornado funnel comes from the forward edge of the cloud, travels at the edge of the cloud downdraft, where it picks up "spin" about a horizontal axis, then is bent upward by the cloud updraft. The stretching by the updraft tightens the vortex, resulting in a tornado. Most Northern Hemisphere tornadoes spiral in a counterclockwise direction, while those in the Southern Hemisphere rotate clockwise. This led some scientists to believe that the Coriolis Effect was somehow involved in the tornado's initiation but this is unlikely since such amplification would take too much time and besides, some Northern Hemisphere tornadoes have been reported to rotate in a clockwise direction.

The true speed within a tornado has never been measured, although it is believed that velocities up to 223 mps (500 mph) are possible. A pioneer in the investigation of tornadoes, T. Theodore Fujita, developed a scale to define its strength based on the damage potential.

NUMBER	WIND SPEED (3 sec. Gust)	DAMAGE
F-0	28–37 mps (65–85mph)	Light
F-1	38–48 mps (86–110 mph)	Moderate
F-2	49–59 mps (111–135 mph)	Considerable
F-3	60–72 mps (136–165 mph)	Severe
F-4	73–88 mps (166–200 mph)	Devastating
F-5	Over 88 mps (over 200 mph)	Incredible

Fujita Wind Damage Scale[11]

Tornadoes are among the most powerful natural phenomena in the world and cause massive destruction wherever they travel along the ground. Whole towns have been ruined; seventy-ton train coaches have been picked up and tossed about like toys; trees have been turned into kindling; houses have been flattened and, of course, people have died. In 1925, one of the most deadly tornadoes followed a course from southeast Missouri to Indiana, a distance of 350 km (219 miles), killing 689 people and injuring almost 2,000, while destroying seven towns in its path. Few people have survived an encounter with a tornado, but one who did, Bill Keller of Kansas reported:

> *On the afternoon of June 22, 1928, between three and four o'clock, I noticed an umbrella-shaped cloud in the west and southwest and from its appearance suspected there was a tornado in it. The air had that peculiar oppressiveness which nearly always precedes the coming of a tornado.*

---

11   The scale actually extends to F12, the speed of sound. Very few tornadoes reach even F5 strength and no recorded tornado has yet surpassed F5 and it is unlikely that one ever will. Note that this is the new enhanced F Scale implemented in 2007

I saw at once that my suspicions were correct. Hanging from the greenish black base of the cloud were three tornadoes. One was perilously near and apparently headed directly for my place....

Two of the tornadoes were some distance away and looked like great ropes dangling from the parent cloud, but the one nearest was shaped more like a funnel, with ragged clouds surrounding it. It appeared larger than the others and occupied the central position, with great cumulus clouds over it.

Steadily the cloud came on, the end gradually rising above the ground. I probably stood there only a few seconds, but was so impressed with the sight it seemed like a long time.

At last the great shaggy end of the funnel hung directly overhead. Everything was still as death. There was a strong, gassy odor, and it seemed as though I could not breathe. There was a screaming, hissing sound coming directly from the end of the funnel. I looked up, and to my astonishment I saw right into the heart of the tornado. There was a circular opening in the center of the funnel, about fifty to one hundred feet in diameter and extending straight upward for a distance of at least half a mile, as best I could judge under the circumstances. The walls of this opening were rotating cloud and the whole was brilliantly lighted with constant flashes of lightning which zigzagged from side to side....[12]

Like a spinning top, a tornado moves erratically, and may hop over an obstruction such as a hill or even a house. In the hills of western Massachusetts, a series of tornadoes touched down on May 29, 1995. The most powerful one struck the town of Egremont at about 7:10 PM and traveled for 24 km (15 mi) churning up hillsides and hopping over the leeward side of hills. Having lived nearby, I saw the extent of its destructive power. In its wake, houses, barns and the grandstands over a fairground were destroyed. For many years later, the course of the twister remained visible by the absence of trees in a path about 300 m (330 yd) wide.

Other Clouds

Clouds, whether scattered or blanketing the sky, appear to the casual observer to be random forms or, at best, one of hundreds or

---

12   A.A. Justice, "Seeing the Inside of a Tornado," *Monthly Weather Review*, May, 1930, p.205

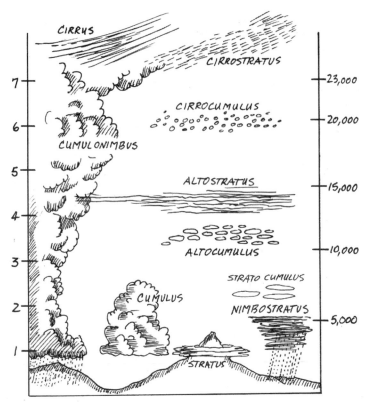

Fig. 5.5 Cloud catalog

thousands of different types. This mysterious, constantly changing vapor was studied as *nephology*[13] even as early as the seventh century BC. Philosopher/scientists such as Aristotle recognized that most clouds are formed by water vapor resulting from the solar heating of surface water, but he did not recognize distinct differences between types of clouds. Sailors were the first to associate different clouds to current and approaching weather:

- ⇛ When clouds appear like rocks and towers, the earth's refreshed with frequent showers;
- ⇛ The higher the clouds, the fairer the weather;
- ⇛ If you see the sun set in a cloud, it will rain tomorrow.

But until 1783, when balloonists first rose into and above the clouds, observations were limited to the view from Earth. The clas-

---

13   Termed 'cloud physics', today.

sification of clouds was therefore also severely limited. An eminent scientist, Robert Hooke, peered high in the sky in 1665 and proposed, "Let Hairy signifie a Sky that hath many small, thin and high Exhalations which resemble locks of hair...." These are the high clouds we now call Cirrus, which do look wispy and hair-like. It was but one of many attempts at classifying cloud formations.

A young chemist, Luke Howard (1772–1864), was fascinated by the ever-changing sky and began to observe cloud formations with the goal of coming up with a means of categorizing cloud types. He was aware of previous work that tended to use words like hairy, broom-like, rock-like, wooly and hazy; and sought more precise definitions that were not dependent on one language. He was also aware that the French naturalist Jean Lamarck had recently proposed a classification based on the height of the clouds: High (more than 7 km), Mid-level (3–7 km) and Low (less than 3 km). Howard, however, proposed to classify clouds by their basic appearance, giving them Latin names to remove any association with one nationality:

- **Cirrus**: mostly ice crystals that appear high in the sky like wisps of hair.
- **Cumulus**: heaped up by rising currents, these clouds appear wooly or puffy.
- **Stratus**: flat clouds that are sheet-like and exhibit no vertical movement.

He further categorized the types that fall into the three height configurations:

- High: **Cirrus, Cirrocumulus, Cirrostratus**
- Mid-level: **Altocumulus, Altostratus, Nimbostratus**[14]
- Low: **Cumulus, Stratus, Stratocumulus**

These, together with the granddaddy of them all, the towering thunderheads or **Cumulonimbus,** comprise the ten categories universally accepted today (Fig. 5.5). Howard also recognized that cloud formations were evanescent and subject to change and transformability. So, when he presented his paper to the Akesian Society in London in December 1802, he named it "On the modifications of clouds." His system was generally accepted, and although others tried to modify or improve upon the general categories, only the

---

14 *Nimbus*, attached to a cloud name, implies that it produces heavy precipitation. *Alto*, means high.

French attempted to use different words (in French, of course) to describe the various cloud types.

Fog is merely a cloud whose base rests at ground level. There are three recognized types:

- **Radiation fog** most often occurs in the summer when, at the end of the day, moist nighttime air near the Earth's surface cools by radiation until condensation forms. It can occur anytime there is sufficient radiative cooling near the ground to cause condensation and enough wind to prevent the condensation from forming dew.

- **Advection fog** occurs most often along the coast as warm air from the ocean moves over cooler shoreline waters and the cool land.

- **Evaporation fog** occurs when water vapor is pumped from a warm ocean or lake into cool moist air that can no longer absorb it. It is like steam rising from a pot of boiling water and is also seen after a heavy rain on a warm forest.

Clouds alone do not define weather, but together with wind, they are guides to current and future weather.

6

The Last Great Adventure
Transglobal Balloon Flight

There are some whole countries where it
    never raines,
Or at least very seldome,
But there is no Countrey where the Winde
    doth not blow,
And that frequently.

Francis Bacon, 1622

Leonardo da Vinci (1452–1519) possessed one of the most inquisitive minds of any human in history. He wanted to understand the world around him and, although he had little formal education, his curiosity led him to study science, engineering, and music as well as art and architecture. It was, therefore, natural that he should ponder the question of air movements, and particularly the power of hot air. About 1480, he devised a roasting device that could be inserted in a fireplace. It consisted of a vaned turbine linked by rods and gears to a rotating spit (Fig. 6.1). The rising hot air drove the turbine, which, in turn, rotated the spit. "The roast will turn slow or fast depending on whether the fire

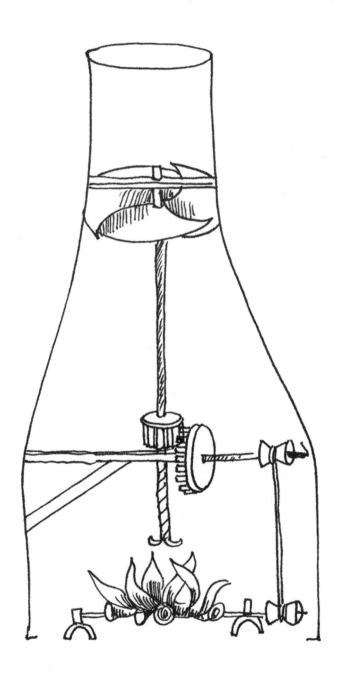

Fig. 6.1 Leonardo's Spit

is small or strong," noted Leonardo. Considering that Leonardo's inventions included a parachute and helicopter-like flying screw, it is surprising that he never applied the principle of rising hot air to the balloon. Had he read the scientific works of the English philosopher Roger Bacon (1214–1292), he might have come across the speculation that man could fly if he were fastened to a large hollow ball of thin copper that was filled with liquid fire or air. Three hundred years passed before other inventors considered the possibility of manned flight.

Joseph Montgolfier, a French papermaker, observed smoke rising from a fire in his fireplace and, like Leonardo, wondered what force caused it to rise. Together with his brother, Étienne, he also noted that clouds were suspended in the atmosphere, and thought that if he could encapsulate the hot air from his fireplace in a bag, it might rise up and float like a cloud. The brothers constructed an 11 m (38 ft) diameter silk bag lined with paper and open at the bottom. In June 1783, they placed it above a charcoal fire. As the bag filled with what they called "Montgolfier gas" (not knowing that it was merely hot air), it rose upward, reaching an altitude of 1 800 m (6,000 ft) and traveled about one and a half kilometer before descending, as the hot air inside the bag cooled. We now know the reason for the balloon's flight: The trapped, lighter, hot air provided the buoyancy or lift to propel the balloon upward by displacing an equal volume of the heavier surrounding air.

Within three months of this first experimental flight, the Montgolfier brothers convinced the French king, Louis XVI, to witness the flight of the first balloon passengers. A sheep, a rooster, and a duck stayed aloft for less than five minutes. Only the rooster suffered any injury; the slight mishap of an injured wing caused by the excited sheep kicking him. Less than a month later, also under the sponsorship of the king, two men, Pilatre de Rozier and Marquis d'Arlandes, were the first aeronauts to fly, ascending in a brightly decorated lighter-than-air craft to the amazement of 400,000 Parisians who turned out to witness the event (Fig. 6.2). Benjamin Franklin, who was in the audience that day, was asked by a bystander, "What good will such a contraption be?" Franklin, who foresaw its military application as an observation platform, answered condescendingly, "What good is a baby?"

Fig. 6.2 Montgolfier's balloon

The need to use fire to heat the air in the balloon proved to be an obvious disadvantage because of the need to carry a heavy brazier and quantities of charcoal or straw fuel in the wicker basket hung from the balloon. The brothers Montgolfier realized this and engaged a young scientist, J.A.C. Charles, to propose a solution. Charles was familiar with the isolation of hydrogen carried out in 1766 by Henry Cavendish, and realized that the use of this lightest of all gases (at one-fourteenth the density of air) could provide greater lift than hot air. But he needed a suitable container other than the paper, which had earlier proved inappropriate because of its permeability to gas. A bag of rubberized silk was chosen and, inflated with hydrogen, rose into the air on August 23, 1783, before the first manned balloon flight. Benjamin Franklin arranged for a demonstration that took place ten years later in Philadelphia. Seeing the balloon drifting high overhead, President George Washington was duly impressed, and thereafter both hydrogen and hot-air balloon flights were carried out for weather and military observation, and later for sport and exploration.

Balloons are subject to the whims of wind and air currents, since they carry no mechanical means of propulsion. Long-distance balloon flights must therefore respect winds and follow a generally easterly route in the temperate latitudes (35–55°) following the prevailing westerlies. This was clear to Jean Pierre Blanchard and Dr. J. Jeffries, who, on Jan 7, 1785, became the first aeronauts to attempt to cross the English Channel. They flew, or rather drifted, from Dover to the French coast in a hot-air balloon. Because they carried no fire, their balloon began to descend when they were about one-third of the way across the channel. The frightened aeronauts began to throw overboard anything that was not attached: anchors, rope, and even pieces of their clothing. Slowly, the balloon began once more to ascend, and, aided by warm air rising from the landmass, it cleared the coast of France and landed in an inland forest. However, an attempt to fly from France to England by Pilatre de Rozier ended in tragedy when his balloon burst into flame at an altitude of 900 m (3,000 ft) and plunged to earth. Not only was he trying to fly against the prevailing winds, but he was also using a double balloon with an upper hydrogen envelope and a lower fire balloon—not appreciating the extreme flammability of hydrogen. Nevertheless, the double-balloon concept, now called the de Rozier type, was actually very clever and has been used in most recent long-distance flights.

In order to fly a gas balloon at a particular altitude, a balance must exist between its weight and the buoyancy of the displaced air. Since the temperature of the air changes with altitude as well as with geographic location, it is difficult to hold a specific altitude without constantly changing the weight of the balloon or the volume of the gas (or hot air). For this reason, J.A.C. Charles introduced a valve at the top of the gas bag to permit gas to be vented, and ballast such as bags of water or sand that are carried aboard the basket and can be thrown overboard. As the balloon's air temperature increases, gas must be vented to reduce buoyancy, and as the temperature drops, ballast must be shed to stop the balloon's descent. It is therefore obvious that long flights need lots of gas and ballast in order to avoid running out of either. This proved to be a formidable obstacle to aeronauts in the nineteenth and early twentieth centuries seeking to make transcontinental flights. Balloons were therefore relegated to exploring the composition of the upper reaches of the atmosphere and to discovering the physiological effects on the body of decreasing concentration of oxygen at higher altitudes. In 1862, a British scientist, James Glaisher, attained an altitude of more than 9000 m (30,000 ft), barely escaping death in the thinner atmosphere because he did not use supplementary oxygen. The incredible height of 34668 m (113,740 ft) was reached by two U.S. Navy officers in 1961. Sadly, one of them, Lt. Cmdr Victor Prather, drowned when his pressure suit filled with water upon landing in the Gulf of Mexico. Long-distance flights, however, awaited new developments in materials and technology.

The years following the Second World War witnessed great technological advances: The propane burner, invented in 1960 by Edward Yost, led to a practical hot-air balloon; space-age materials such as *mylar* led to stronger and lighter gas bags. Helium, an inert gas that is only slightly heavier than hydrogen, was isolated at the turn of the twentieth century and provided a safe alternative to flammable hydrogen. Extremely rare, it was too expensive for everyday use in balloons, but has, nonetheless, proven to be essential in facilitating long-distance balloon flights.

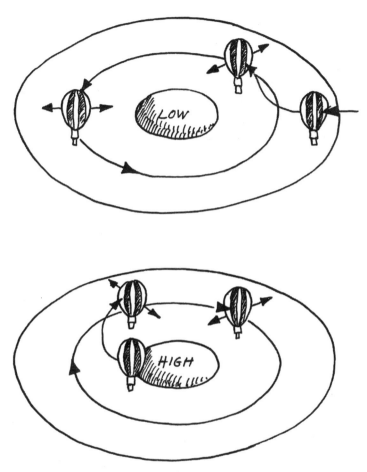

Fig. 6.3 Coriolis Effect on balloon's path

Where the Wind Blows

If a balloon simply drifts with the wind, then how can the adventurous aeronaut navigate and aim for a particular destination? To answer this question, we must re-examine the flow of the wind constantly taking place between zones of high and low pressure. Were it not for the Coriolis Effect, wind, and our balloon, would flow directly from a high-pressure area to a low-pressure area. The Coriolis Effect, however, turns it to the right (in the Northern Hemisphere). The result is that the balloon will tend to move clockwise around a high-pressure area and counterclockwise around a low-pressure area, always seeking to balance the pressure gradient force with the Coriolis force (Fig. 6.3). Our balloon would be doomed to wobble around forever about either of these areas, were it not for the fronts that divide regions of high and low pressure. These fronts are themselves also constantly moving in a generally easterly direction in the Northern Hemisphere and define boundaries along which there are disturbances such as storms. Typically, a cold front slides under warmer air ahead of it, and a warm front climbs over the colder air ahead of it (Fig. 6.4). There are many variations of these fronts that exist in nature but for our purpose we will examine only what happens along typical fronts, to illustrate some of the options that are open to our aeronaut.

Along a warm front, since the warm air is rising, our balloonist would rise as well. There could be rain ahead of the front, but the weather would generally be stable. By contrast, along a cold front there could be some strong updrafts of warm air and our balloonist might spot thunderstorms, which he would definitely want to avoid. Ahead of the front, there may be a low-level air jet that could propel him very nicely along the frontal boundary (Fig. 6.5). If our

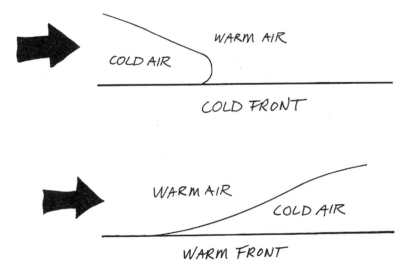

Fig. 6.4 Warm and Cold Fronts

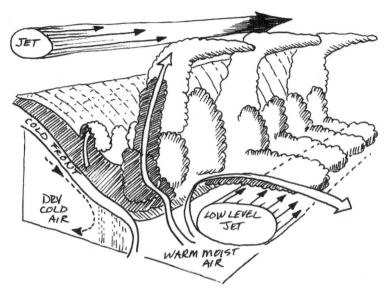

Fig. 6.5 Air flow around frontal boundaries

balloonist were to climb to around 9 000 m (30,000 ft), he could encounter the jet stream that follows a path between the colder arctic
air and the warmer tropical air. This region, with winds in the core
of up to 85 m/s (190 mph), could quickly propel our intrepid traveler
around the world, provided he could stay in its ever-changing track.
Of course, our balloonist would be advised to travel in the winter when the jet stream moves south (in the Northern Hemisphere)
rather than in the summer when the jet stream is far to the north
and the overall weather is more unstable, with possible thunderstorms and even hurricanes in his path. Our balloonist should also
be armed with the knowledge that his balloon may lose altitude at
night when the air temperature drops, as well as over water, which
cools more rapidly than the landmass. The balloonist approaching
the Rocky Mountains, the Alps, or the Himalayas knows that winds
blowing toward them are deflected upward, creating a wave that
will propel him up and over the mountains. Such waves can extend
a great distance downwind of the mountain and could serve our
balloonist well by speeding him on his journey, as well as giving
him a bumpy ride because of strong turbulence.

Race For The Record

By 1996, balloons had been flown across the Atlantic and Pacific oceans and all of the world's continents. There remained but one challenge, a flight around the world. Suggestive of the prize that was awarded for the first flight in an airplane across the Atlantic, a million-dollar prize awaited the first team to successfully circumnavigate the globe in a balloon.

In the winter of 1996, five teams prepared for the challenge. Three of the teams planned flights above 10 500 m (35,000 ft) and two, no higher than 8 400 m (28,000 ft). The first to attempt a take-off was the *Virgin Global Challenger* in Marrakech, Morocco. While setting it up on December 9, 1997, this balloon broke free, ending the attempt. The second challenger, Kevin Uliassi, took off on December 31, 1997 from Rockford, Illinois, in the *J. Renee*, a 54 m-high balloon that was forced to land in Indiana within a short time because of equipment failure. On the same day, the third challenger, Steve Fossett in the *Solo Spirit*, left St. Louis, Missouri. Two days later, he reached London, and the following day flew over Budapest. The following morning, however, equipment failure forced Fossett to land in Krasnodar, Russia. He achieved only slightly more than half the distance he had a year earlier, when he reached India.

On January 9, 1998, Bert Rutan, who had found fame by flying around the world in a plane, took off with his copilot in the *Global Hilton* from Albuquerque, New Mexico. Upon reaching 8 480 m (27,500 ft), the helium cell burst, dooming the flight. The two pilots parachuted to earth while the balloon continued its flight with the remaining helium, finally landing in northern Texas. The Breitling Orbiter made the final attempt of the season on January 28. Launched from Chateau d'Oex in the Swiss Alps, the vehicle was manned by three balloonists, including Bertrand Piccard, the grandson of Auguste Piccard. Auguste invented both the pressurized stratospheric balloon, with which he was the first to ascend to more than 17 000 m (55,000 ft), and the deep-sea diving bathyscaph in which his son, Jacques, descended more than 10 500 m (35,000 ft) into the Marianas Trench.

The Orbiter team was plagued by lack of wind speed in the first part of their journey and only entered the jet stream over Syria, three days after departure. At this juncture of the flight, the crew was awaiting clearance to cross China where a rapid jet stream would carry them at 230 km/hr (144 mph) to the coast of California. As they crossed India, permission had still not been granted, so they descended to a lower level to slow the flight. Barely moving, the balloon hovered 40 m over the Indus River to the delight of crowds of children on the banks. When the crew finally received permission from the Chinese authorities on February 6, it was too late to catch the jet stream. They landed the next day on the road to Mandalay in Myanmar (Burma), satisfied that they had flown longer than any other craft without re-fueling, 9 days and 18 hours.

1998–1999, The Final Season

Not to be undone by his previous failure, Fossett made his next solo attempt starting from Mendoza, Argentina, in August 1998, the first such attempt in the Southern Hemisphere. The 24000 km trip was expected to take 18 days. Under a full moon, the balloon lifted off ahead of schedule. After ten days and 15000 km (9,300 mi), Fossett encountered a severe thunderstorm off the eastern coast of Australia. Lightning struck his balloon while he was at 7000 m (23,000 ft), and he began a rapid descent to the shark-infested Pacific Ocean. Shaken but not injured, Fossett was rescued and soon started planning his fifth attempt to complete the last great adventure.

In a race reminiscent of that preceding Charles Lindbergh's successful solo airplane Atlantic crossing in 1926, the goal of circumnavigating the globe by balloon drove competing teams to seek the edge that would lead them to achieve victory. Unlike the wicker basket that was suspended from the primitive early balloons, the challengers used space-age materials for both the balloon envelopes and the gondolas.

The RE/MAX team, led by Dave Liniger, planned an attempt using a balloon the height of the Empire State Building with a spacecraft-like gondola, set to fly at altitudes between 24000 m (80,000 ft) and 39000 m (130,000 ft), far above any weather. The craft was to lift off from Australia in December 1998, but because of poor weather the flight was postponed for a year. Richard Branson invited Steve Fossett to join him and Per Lindstrand in a year-end flight of the ICO Global Balloon. After crossing nineteen countries following a perfect launch from Marrakech, Morocco, the balloon was propelled by the jet stream across the Pacific Ocean at speeds of up to 320 km/hr (200 mph). However, a deepening trough in the central Pacific deflected the balloon away from the jet stream, slowing it and dooming the flight. "We've been beaten by the weather," said the project director, as the balloon was led to a sea landing on Christmas day, north of Honolulu, Hawaii, and 3 200 km (2,000 mi) short of Steve Fossett's solo record of the previous August.

On Feb 17, 1999, the Cable & Wireless Balloon team took off from Almeria, Spain, and headed south to catch the jet stream they hoped would propel them around the globe. It was the 24th attempted circumnavigation by balloon since 1980. On March 1, Bertrand Piccard's birthday, the Breitling Orbiter 3, with Brian Jones as co-pilot, took off from Chateau d'Oex, Switzerland, heading for Italy. The chase for the prize was on! Apart from the weather, the two flights were now at the mercy of both geographic and political obstacles.

The Cable & Wireless team entered the northern sub-tropical jet stream three days after takeoff. But it lost precious time, having to thread its way south to a 48 km (30 mi) wide wind corridor that would keep the flight clear of Chinese air space, since it had not been granted flyover permission. The balloon crossed the coast of India on the tenth day of the flight with more than one third of the anticipated flight completed. On the day of the Orbiter's takeoff, the C&W balloon was over the Bay of Bengal. On March 6th it reached the southern coast of Japan, having successfully avoided coming near Chinese airspace, and headed toward Sapporo to enter the northern branch of the fickle jet stream that had split into three tracks. By then, Orbiter had crossed the northern boundary of Yemen, after some difficulty when they were at first refused permission to fly over Yemen's northern military zone. Clearly, the Cable & Wireless Balloon was in the lead with no apparent opportunity for the Orbiter to catch up. But...

On March 7th a low-level trough deepened southeast of Tokyo, right in the path of the leader. Flying under clouds, the C&W balloon tried to rise through them to escape the weather and recharge its batteries, which were needed to run the kerosene pumps. Instead, the balloon was being pushed eastward, away from its northern track. Having survived the freezing of a heat pump over Saudi Arabia, being dragged toward a high-pressure cell over the Arabian Sea, and a wild ride around thunderstorms over Thailand, the aeronauts, Colin Prescot and Andy Elson, were forced to abandon their

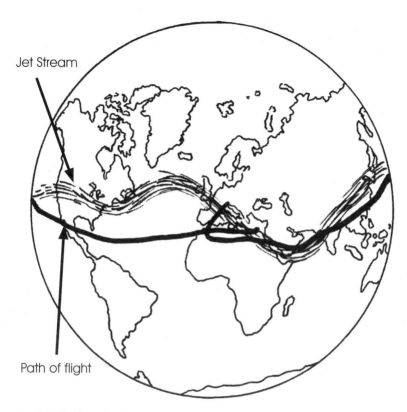

Jet Stream

Path of flight

Fig. 6.6 Path of the balloon circling the earth

flight, ditching the Cable & Wireless balloon in the Pacific Ocean off Japan.

With its competitor defeated by a wall of clouds, the Breitling team still had a long way to go before reaching its goal. Since the balloon was of Swiss registry, China permitted the Breitling team to fly over the country as long as the balloon stayed below the 26th parallel. This permitted it to reach Japan six days faster than it had taken the C&W team. The team members now faced the decision of which jet stream to take: the polar jet stream that would direct them toward the coast of the United States, or the sub-tropical jet stream that would lead them across Mexico. Since it is more difficult to cross the United States with its heavy air traffic, the team headed for Mexico. Skirting thunderstorms, they passed near the point south of Hawaii where Branson's flight had ended. Nestled in the jet

stream, they had broken Steve Fossett's distance record by March
14th. Popping out of the jet stream over Mexico, they were suddenly
pushed by winds southward toward Venezuela. Fortunately, meteorologists found an altitude that turned them eastward again and
into a 160 km/hr (100mph) jet stream over the Atlantic Ocean on
the last leg of their journey. On the 19th, they broke the endurance
record set only twelve days earlier by Elson and Prescot and on the
next day they passed victorious across longitude 9°–27°W, completing their circumnavigation of the globe in Mauritania (Fig. 6.6).
They continued their flight for another day, landing in Egypt after a
46 400 km (29,000 mi) journey of twenty days, one quarter the time
it took Phineas Fogg to complete his adventure in Jules Verne's tale,
Around the World in 80 Days.

7

Transformations
Clouds Shed Tears

Colder and louder blew the wind, a gale
from the Northeast.
The snow fell hissing in the brine, and the
billows frothed like yeast,
Down came the storm, and smote amain,
the vessel in its strength.

The wreck of the Hesperus,

Henry Wadsworth Longfellow

Cycles

Without water, life would be impossible. In its liquid state, it nourishes the earth's soil, flows down rivers to replenish the oceans, and cools and cleans our bodies (which are, themselves, 70% water). In its gaseous form, water vapor fills the atmosphere in various concentrations (see p. 39–40), making us feel either dry or sweaty; condenses into water droplets to become clouds; or cools into ice crystals that may aggregate into snow. As a solid, ice covers the earth's two poles and has at times spread to wrap large parts of the world in a frozen blanket. These three states, which reveal the nature of water, represent a transformation that is constantly taking place. Water evaporates from oceans, lakes and streams; forms clouds that travel about and may drop their condensates, rain or snow, on land or sea; and forms ice, either through freezing of water or compacting of snow. Whether as rain, snow or ice, water returns to the oceans, lakes, or streams to start the cycle over again.

To change water from one state to another, heat energy is involved. In water's frozen state, its molecules move about very slowly, expending very little energy. By contrast, in water's liquid state, its molecules move about much more rapidly and therefore have more heat energy. Consequently, to change ice to water, a great deal of heat energy must be introduced. In fact, this change of state requires the same amount of heat energy as is needed to raise the temperature of water from 0° to 80°C. Yet, the temperature of the water remains constant because the energy is used to loosen the molecular bonds allowing the molecules to move past each other and speed up rather than raising their temperature. Incidentally, speaking of heat, it does not exist. It is merely a term given to our observation. When we feel a rise in temperature, we are in fact feeling an increase in the speed of molecules around us.

When water turns to vapor, as much as seven times more heat energy is needed, since the water molecules in vapor move around more rapidly than in water. However, these same water molecules are spread far apart in vapor, so the temperature of the vapor may not necessarily change with the application of heat energy: The molecules may simply move faster and spread farther apart. Since heat energy is needed to **evapor**ate water, the temperature of the water must decrease. We experience this cooling effect, for instance, when perspiration evaporates from our skin or water evaporates from a wet bathing suit.

Rain

Following the cyclical process one step further, water vapor *condenses* into water droplets such as dew (see p. 40) or microscopic droplets that are seen as clouds. In the process, heat energy is released, beginning the process of forming clouds and producing rain. As warm, moist air rises, it expands and is consequently cooled by the decrease in atmospheric pressure with increasing height (at the rate of 1°C / 100 m [5.4°F / 1,000 ft][15]. A level is reached where the cooler air becomes fully saturated and water droplets begin to condense around microscopic particles of dust. This aggregation of microscopic water droplets, about one hundredth the size of an average 2-mm raindrop, becomes visible to us as clouds[16]. In the process, heat is released to the surrounding air, causing the newly formed clouds to churn and rise faster. It would take a great deal of time for raindrops to form by condensation alone. But, if the air temperature in the cloud is above about −10°C (14°F), water droplets in the cloud move about and collide with each other, forming larger droplets by **accretion** that eventually reach the size of drizzle that falls to earth. Unlike liquid water, small water droplets in clouds will not freeze immediately at 0°C and in fact may remain as a **supercooled** liquid in storm clouds until reaching a temperature of about −37°C (−34°F). This supercooled water is "metastable," and will remain liquid until it freezes around a nucleus such as a dust particle or ice crystal, which generally takes place at about −10°C (14°F). In clouds high in the atmosphere, ice crystals grow faster than water droplets, and rain will always start as ice crystals. In summer, these turn to water as they pass through the warmer lower atmosphere.

Snow

In a cloud that reaches an altitude where the temperature is less than between −15°C (5°F) to −20°C (−4°F), water droplets primarily form ice crystals that grow very quickly as the surrounding water droplets attach themselves to the newly formed crystals. As these crystals grow larger and heavier, they will eventually start to fall through warmer air below, slowing the crystals' growth and forming snowflakes with a characteristic hexagonal shape. If the lower-level air temperature is below freezing, snow will fall to the ground, but if it is above freezing, the snowflakes will melt and reach the ground as rain (for reference, 250 mm [10 in] of fresh dry snow is equivalent to 25 mm [1 in] of water).

---

15   During expansion, air does work and hence loses energy.

16   The ratio of the size of a cloud-drop to a raindrop is roughly the same as the ratio of the size of the earth to the sun (1:100 in diameter; 1:1,000,000 in volume).

Snowflakes

For his fifteenth birthday in 1880, a young Vermont farm boy, Wilson Bentley, received a microscope. Swaddling himself in multiple layers of clothing to ward off the numbing cold, he peered through the lens at the snowflakes that landed on glass slides. He was fascinated with the seemingly infinite variety of snow crystals and began photographing them. With immeasurable patience, he took over 4,500 photo micrographs of frost, dew, and snow crystals, half of which were published in 1932 in a seminal work, *Snow Crystals*. The "wonderfully delicate and exquisite figures," he wrote, reveal "the history of each crystal and the changes through which it has passed in its journey through cloudland." By observing their form and structure, he was able to determine the origins of crystals in different cloud formations and the air temperature at the moment of crystallization. Bentley also studied the size of raindrops, determining that the largest was about 6 mm (¼ in) in diameter and that its size could be related to the size of snowflakes in clouds, the type of storm, and even the presence of lightning.

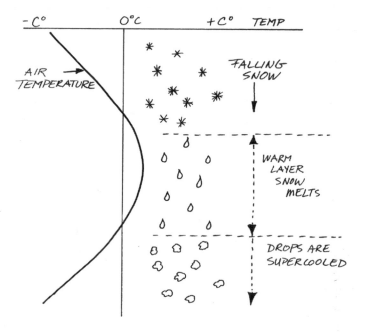

Fig. 7.1 Formation of freezing rain

Ice

As every pilot knows, the greatest danger to the stability of an airplane is the accumulation of ice, particularly on the wings. Clear ice will accrete on the front edge of a wing passing through cumulus clouds associated with lake or marine air masses, with temperatures in the 0° to –10°C (32° to 14°F) range that therefore contain super-cooled water. Rime ice, which looks milky and opaque, is associated with passage through stratus clouds in the –10° to –20°C (14° to –4° F) range, which contain smaller water droplets. Freezing rain will oc-cur when snow drops through a warm layer into a colder layer below, as can happen along the boundary of a warm front (Fig 7.1, see also Fig 6.4). When striking the cold surface of a wing, these supercooled wa-ter droplets freeze instantly and can quickly accumulate to substan-tial thicknesses. Ice has been implicated as the cause of numerous air-plane crashes, although rarely to high-flying jets that spend little time flying through clouds in the dangerous temperature ranges. Since ice that affects flight occurs in or near clouds, the size of associated wa-ter droplets is quite unlike what we are used to seeing falling as rain. Most water droplets that result in icing are in the 15 to 40 micron range, much smaller than the dot at the end of this sentence.

Frost is the cold-weather equivalent of summer dew and will form on surfaces on cold clear nights when, at sunset, the tempera-ture and dew points are within about 3°C (5°F) of each other. Typi-cally, water vapor density is highest near the ground, increasing the chances of temperature falling to the frostpoint at the surface to produce frost. Frost also forms on the inside of a cold window where the interior humidity is high.

Hail

When the atmosphere near the surface is warm and moist com-pared to the cold air above, rising air remains warmer than the air through which it passes. This leads to an unstable atmosphere and encourages the growth of the massive clouds known as cumulonim-bus or thunderstorm clouds. Within these clouds, violent convective currents cause updrafts and downdrafts that can trap ice crystals, keeping them suspended and in constant agitated motion. As they are tossed about, these crystals are pelted by supercooled droplets of water that freeze instantly, adding a milky rime or a smooth glaze to the crystal. Caught in a powerful updraft, the enlarged crystal, now a **hailstone**, is hurled back toward the top of the cloud, only to start its descent again, enlarging its mass. This cycle is repeated until the hailstone reaches a weight that can no longer be held up by the force of the updraft. Some hailstones actually keep traveling through different parts of the cloud without ever falling to earth. Hailstones that do fall to earth can be the size of small pearls or as large as oranges.

The Moon Has A Golden Halo

A strong pioneering spirit in the young United States enticed many families to move west, seeking new opportunities, free land, and a better future. Among these, in 1846, were the families of two prosperous Illinois farmers. They had heard of the sunny, warm land of California that, although still a Mexican state, was expected soon to become a United States possession (following the war with Mexico, 1846–1848). Neither the sixty-two-year-old George Donner, nor his brother, Jacob, was qualified to lead a party of eighty-seven people that would include farmers, many women and children, and a number of elderly relatives. As plains dwellers, they had never experienced life in the wilderness nor the severity of mountain weather. Nevertheless, they set out from Springfield, Illinois, in April 1846, joining other California-bound pioneers, and drove their wagons west, reaching South Pass in southwestern Wyoming in early summer. There, the Donners and several other families were convinced by Lansford Hastings to try a new route to California that was supposed to save 560 km (350 mi), and that closely follows the trace of today's US Interstate Route 80. Hastings was a promoter who had never actually traveled the route. In fact, in the summer of 1846, the route was barely passable with a difficult crossing of the Wasatch Mountains (where the Donner party lost its way) and a hot and dry crossing of the 128 km-(80 mi)-wide Great Salt Lake desert. Before rejoining the main route, along the Humboldt River valley, the party had actually traveled 200 km (125 mi) longer than the conventional trail, lost many of its oxen and cattle, and was three weeks behind the main wagon train that had taken the normal route (Fig. 7.2). The exhausted party reached the foothills of the Sierras, west of Reno, in late October, and decided to rest before continuing along the Truckee River and undertaking the hazardous mountain crossing. This proved to be a tragic error.

As the pioneers gathered their strength, a low-pressure area that had formed in the northwestern United States was also gathering

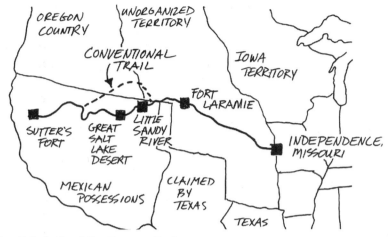

Fig. 7.2 Path of the Donner Party

strength. The counterclockwise circulation around the low pulled in moisture from the Pacific Ocean, forming clouds along the cold front that moved toward the Sierra Nevada Mountains. Rain fell in the Central Valley region of California, where the temperature was still moderate, but as the ground rose toward the mountains, the circulating air and clouds were forced upward. With every foot of height, the ground temperature dropped, and it reached the freezing level at about 1500 m (5,000 ft) where the precipitation released from the clouds no longer turned to rain but remained in its snowy state. It is not unusual for the first snows of the season to drop in October: Two years earlier, the Stephens Party crossed the snowy mountains around the end of November, but had the advantage of being led by experienced mountain men. What was unusual in the winter of 1846 was that the first snows were followed within a week by a second storm that caused the 2200 m- (7,200 ft)-high pass to be covered with about 2.4 m (8 ft) of snow. This was the first of a total of 7.2 m (24 ft) of snow estimated to have fallen that winter, far from the record 20.4m (68 ft) that fell in the winter of 1931–2, but enough to precipitate tragic events.

The two storms at the end of October that choked the route with more than 2½ m (8 ft) of snow forced the party to set up camp at Alder Creek—8 km (5 mi) from what is today called Lake Donner—in the vain hope that a break in the weather would permit them to cross the pass. But the snow was too deep and did not melt, even when the sun shone brightly. A golden halo around the moon on the night of October 27 warned the party of the second approaching storm. By the next morning it had begun to snow and it was too late to attempt the crossing. The party's fate was sealed by being one day late.

In a brave attempt to reach rescuers at Sutter's Fort (Sacramento), one group of fifteen members of the party set out on foot in mid-December using makeshift snowshoes. A month later, only seven, including all the women and two men, reached the first ranch on the western slope of the Sierras.

At the camp, the survivors, lacking food, had resorted to cannibalism, seeking precious nourishment by eating their dead companions.

Rescue efforts, to bring out the remaining party, were mounted from Fort Sutter during the winter, but were hampered by ten major storms that occurred between October 16, 1846 and April 3, 1847. The organization of relief efforts was also hindered by the war with Mexico that engaged most able-bodied men in California, as well as by the weather. The first relief party reached Donner Lake in late February, and by the time the last survivor was rescued in April, 42 of the original party of 89 had perished, with quite a few providing food for the survivors. Among the dead was the wagonmaster, George Donner, and the last to die was his wife, Tamsen, who was partly eaten by the last of the rescued survivors.

The route the party took became the route of the first transcontinental railroad, which was begun in 1863 and completed six years later with the driving of the symbolic golden spike at Promontory, Utah. However, even today, the railroad and the adjacent Interstate highway are sometimes forced to close because of snow accumulations that cannot be plowed aside.

Mountain weather is quite different from that encountered in plains, as the flatlanders of the Donner party discovered. Wind blowing across a plain continues unimpeded, at least until it encounters a pressure boundary. The barrier presented by a mountain changes not only the wind's direction but its temperature as well. Unstable and rapidly changing weather conditions are characteristic of mountain weather. This proved disastrous to the pioneers of the Donner party, but to more competent explorers and mountaineers such uncertainties present simply another challenge.

8

Conquering the Top of the World
Snow, Blizzards and Avalanches

I am the daughter of Earth and Water
And the nursling of the Sky;
I pass through pores of the ocean and
   shores;
I change, but I cannot die

The Cloud, Percy Bysshe Shelley, (1820)

The world's major mountain ranges—the Andes, Alps, Rockies and Himalayas—were formed when one tectonic plate pressed against another, with consequent upward thrusting and folding of the earth's crust. Between twenty-five and thirty-five million years ago, such a movement took place between the African plate and the Eurasian Plate. Along an arc from the Riviera on the Mediterranean coast describing the French-Italian and Swiss-Italian borders and terminating on the border of Slovenia, the Alps rose with snow-capped crystalline peaks thrust up as high as 4807 m (15,771 ft). The mountains stood as a barrier between the Italian peninsula and the rest of Europe.

The amazing feat of the Carthaginian General Hannibal, crossing the Alps near what is now known as the Little St. Bernard Pass on the French-Italian border, was one of the few excursions of men to the high mountains. Yet, as we shall see, even he, with his small army and a train of baggage-carrying elephants on his way to invade Italy, was almost defeated by the mountain. People who lived at the base of the Alps rarely ventured up the slopes out of fear of superstitious legend. The inquisitive Leonardo da Vinci was not deterred by such irrational beliefs and, in 1500, climbed high up the slopes of the Pennine (western) Alps to make meteorological observations. Apart from such exceptions, until the end of the eighteenth century the mountains were only occasionally visited by intrepid pioneers seeking crystals or hunting chamois.

Mountaineers

Eighty km (50 mi) southeast of Geneva, Switzerland, where Italy, France and Switzerland meet, stands the highest and proudest of the Alpine peaks, the Mont Blanc. Ancient trade routes passed within view of its summit, and in the eleventh century a priory was established at its base in Chamonix. Although many viewed its peak with awe, no one ventured up its slopes. Horace Bénédict de Saussure, a naturalist from Geneva who was interested in geology, set out in 1760 on his first visit to Chamonix to study glaciers. "I must get to the top," he wrote when he first saw the mountain. The longer he spent around its base, the more he kept looking up. "It became for me a kind of illness. I could not even look upon the mountain, which is visible from so many points round about, without being seized with an aching of desire."

In the following decades, a number of expeditions were mounted to reach the summit, including at least one by de Saussure. All failed to reach their goal until August 8, 1786 when a physician, Michel Gabriel Paccard, and a guide and crystal-hunter, Jacques Balmat, cautiously traversed glaciers and slowly climbed up a precarious snow-covered ridge before successfully reaching the summit, exhausted but victorious. As the driving force behind this success, it was de Saussure who became recognized as the man responsible for wiping out the superstition with which high places were regarded.

The glaciers, those mysterious rivers of ice, attracted the first scientist-mountaineers, like de Saussure, to climb the Alps. Based on his own observations, Louis Agassiz, the Swiss-American naturalist, first presented a theory, in 1840, describing the growth and movement of glaciers. He believed that Pleistocene glaciation (between 2½ million and 11,000 years ago), when the polar glaciers spread to the temperate regions, was caused by progressive cooling of the atmosphere. Since that time, we have learned that there was not just one glacial period, but many, and their cause remains a matter of debate among scientists. Some of the theories that have been advanced include the possibility of a variation in the tilt or orbit of the earth[17], a decrease in the sun's radiation, an increase in volcanic activity or an asteroid strike throwing up dust clouds that cut off solar radiation, or a decrease in the level of carbon dioxide in the atmosphere. None of these theories is yet universally accepted, although the Little Ice Age (1430–1850) during which the world's average temperature was about 1°C (1.8°F) cooler, is credibly associated with a drop in sunspots (see p. 14). This cooling promoted the growth of alpine glaciers and clearly demonstrated the relation between glaciations and temperature. Benjamin Franklin, while residing in Paris in 1783 as the American ambassador, observed the cooling effect of a volcanic eruption in Iceland. It resulted in an abysmal Parisian winter. Currently, throughout the world, glaciers are retreating, probably as a result of an increase in carbon dioxide in the atmosphere resulting in global warming. For instance, the Zepu glacier in Tibet has lost more than one hundred meters in thickness; the glacier on Mt Kilimanjaro is expected to disappear within a decade. In the 1990s alone, glaciers worldwide shrunk more than 4%, on top of a 10% decrease since 1850.

Blue Ice

Glaciers are made of snow that has been compacted into a dense clear granular ice in a process that may take from ten to one hundred years. Starting as fresh snow on the névé, the fields that lie above the snow line (2400–3000 m [8,000–10,000 ft]), snow crystals are first compressed by the weight of new snow falling above them, becoming granular and as much as ten times as dense as fresh snow[18] but no more than half the density of water. Then, this granular snow, known as **firn** (last year's snow), recrystalizes

---

17   Climatic cycles of 20,000–100,000 years, known as the Milankovic Cycles, are attributed to changes in the orbital characteristics of the earth.

18   The density of fresh snow varies from 20 to 600 kg/m³.

to form linked ice crystals interspersed with air bubbles. Finally, the pressure of the glacier squeezes out most of the air, leaving ice with a density about 90% that of water, and, in the words of Robert P. Sharp, a Cal Tech glaciologist, "a gorgeous, brilliant, deep shade of clearest blue."

Since glaciers grow on sloped mountainsides, they are constantly on the move, propelled by gravity. To explore how glaciers move, Agassiz planted a straight line of stakes from one side to the other of a glacial valley. Coming back the following year, he discovered that the stakes now described an arc pointing down the valley (Fig. 8.1). Apparently, the center surface of the glacier was moving faster than its sides. Further observations determined that a similar phenomenon was taking place in the depth of the glacier, with the top moving faster than the bottom. He concluded that friction between the ice and the rock faces was slowing the movement.

As a glacier creeps down a valley, its surface is pulled apart by the non-uniform movement taking place within the body of its mass. Due to this strain, when the strength of the ice is exceeded, the surface tears apart, opening a **crevasse** that may extend down as much

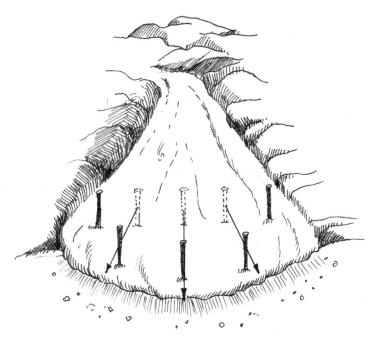

Fig. 8.1 Agassiz's experiment

as 30 m (100 ft). Crevasses can be either longitudinal or transverse to the direction of the glacier's movement, and are among the most dangerous obstacles to mountaineers because weak snow bridges can camouflage them. This was only one of the many obstacles facing a 25-year-old English artist and illustrator who undertook to climb the most famous of Alpine peaks, the Matterhorn.

The Matterhorn

Edward Whymper was commissioned in 1860 to do a series of sketches of Alpine scenery and began an extensive journey throughout the region. He was struck by the majestic beauty of the rocky crystalline pyramid, with its four flintstone-sharp corners, jutting 3 000 m (10,000 ft) above a skirt of glaciers. The Matterhorn, at 4 478 m (14,691 ft) high, is the second highest Alpine peak and serves as a dramatic backdrop for the Swiss resort of Zermatt. On the Italian side it is called Monte Cervino and can be reached from Breuil-Cervina. For a long time, Whymper studied the mountain, plotting how he might be able to climb to its windy summit. It looked foreboding, but in 1861 he made his first attempt, climbing with a guide. The first night, as the two were sleeping, a shower of rocks descended upon them and they barely escaped calamity by finding refuge under overhanging rocks. The second day, when they reached a point known as the Chimney, a crack between two vertical rock faces, the guide refused to go on and they had to turn back. For the next four years, Whymper made six more attempts to conquer "that awful mountain," that were all doomed to failure. On the second attempt, one of the guides fell while crossing the Glacier du Lion and a violent storm struck. On the third attempt, a guide fell ill when they were two thirds of the way to the top. Out of frustration, Whymper then attempted to climb alone, but had to turn around and slipped on his way down, stopping 3 m (10 ft) short of a precipice and certain death. The fifth time, the climbing party was turned back by a vicious storm that turned out to be of short duration—as evidenced by the rapid clearing after they had turned back—but it was too late to move forward again. The sixth time, he could not advance beyond an unclimbable razor-edged ridge. The seventh time he was stopped by a sudden blizzard that started as the sky blackened and cold air rushed down the slope. As lightning struck and thunder raged, the party of six was pinned down for 26 hours. When they returned, exhausted, to their starting point in Breuil, the innkeeper asked, "What storm? It was beautiful all day except for that little cloud on the mountain."

Utterly frustrated, Whymper now decided to risk his eighth attempt following the Northeast ridge instead of the southwest ridge that he had been pursuing. Sadly, this attempt failed when his party was showered by an avalanche of rocks while going up a gully leading to the ridge. He was now in a race against time, as a competing Italian team, led by his former guide, Jean-Antoine Carrel, threatened to steal his glory. Convinced that the east face offered a better path because the natural dip of the rock strata from southwest to northeast would create a natural staircase, he assembled a team of seven climbers and started out on a brilliant clear day, July 14, 1865. The climb turned out to be effortless and the party reached the summit easily, well ahead of the Italian party, who turned back when they saw they had lost the race. Whymper and his party stayed for an hour to drink in the spectacular view with white alpine peaks visible for over 160 km (100 mi). At 2:40 they began their descent. In a matter of minutes they reached a short difficult section with ice-covered rocks. Roping themselves together, they moved cautiously, one man at a time. Suddenly, the second man slipped and bumped into the first, sending them both flying down. The third and fourth men were pulled off and dragged violently off their steps. As the rope whipped about and went taut, Whymper and the other two men braced themselves. There was a violent jerk and the hemp rope tore. "For a few seconds we saw our unfortunate companions sliding downwards on their backs, and spreading out their hands, endeavoring to save themselves. They passed from our sight uninjured, disappeared one by one, and fell from precipice to precipice on the Matterhorn gletscher (glacier) below, a distance of nearly 2 400 m (7,800 ft) in height." One of the victims of the disaster, Lord Francis Douglas, was never found. So ended the spectacular and tragic first climb of the Matterhorn.

Avalanche

As snow falls on a mountain face, adding layer upon layer of fresh snow, a fragile bond between ice particles holds the mass together. If the slope is steep, the bond may be insufficient to keep it clinging to the mountain. A sudden gust of wind, a loud noise, or an inadvertent movement by a climber or skier can cause the snow mass to be released in an **avalanche**. Moving at great speed and with a roar, as it pushes the air out in front of it, an avalanche is one of the most frightening events that a mountaineer can face. The cascading snow that can move at speed of over 300 km/hr (200 mph) will gather up everything in its path, such as rocks, houses or people. Escape is impossible, as the advancing avalanche of snow and debris is lifted off the ground by trapped air along its leading edge.

Hannibal faced this danger as he crossed the Alps in 218 BC with 38,000 men, 8,000 horses, and 37 elephants, at the setting of the Pleiades (about October). At the time a blanket of fresh snow covered crusted snow, a condition indicating potential avalanche conditions. During the descent, animals' hooves perforated the upper fluff and broke through the crust, causing the upper layer to give way and sweeping men and animals down the slope. Half of the men, one quarter of the horses, and several elephants died, but with the remainder of his army, Hannibal fought all the way down the Italian peninsula to the gates of Rome. Avalanches could not defeat him, and only lack of support from his home government in Carthage caused him to eventually retreat.

Two thousand years later, during the First World War, as Austrian and Italian troops were fighting in the Tyrolean Alps, an avalanche killed 253 soldiers on December 13, 1916. Over the next two winters, both sides resorted to firing artillery shells to destabilize the snow hanging high up the mountain over the enemy troops. The resulting avalanches caused the deaths of over 40,000 soldiers, swept up in flying snow and rocks. For years thereafter, frozen corpses emerged from the bases of the mountains in the spring thaw.

Somewhere in the world, avalanches continue to claim new victims every winter.

As a testament to the preserving property of ice, in 1991, the corpse of a European was discovered in the Alps at the base of an ice flow. The man, a Bronze-Age hunter, died 5,300 years ago from a wound in his chest caused by a flint-tipped arrow. The well-preserved body yielded DNA evidence that indicated the fifty-year-old man had descendants in England and Ireland. Perhaps Lord Douglas will appear to some future generation, presented at the base of the Matterhorn's ever-moving glacier.

Glaciers Today

Throughout the world, glaciers are retreating: In Peru, the Quelccaya ice cap, the largest in the tropics, is retreating by 180 m (600 ft) per year and will be gone by the year 2100. The 11,700 year old Kilimanjaro ice cap in Africa has shrunk by 75% in the past century and is projected to disappear completely by 2015. In Tibet, the Zepu glacier that took a century to shrink by 7% until 1960, has taken only forty more years to shrink by another 7%. The Unteraar glacier in Switzerland has receded more than 1.6 km (1 mile) since the mid nineteenth century. All of these changes point to a shift in world weather that has dramatically accelerated since the dawn of the industrial age. The water descending from these glaciers nourishes vast areas of the world, so their disappearance will have severe consequences, with loss of irrigation water as well as drinking water, seriously disrupting the ecology of these regions.

9

The Unsinkable *Titanic*
Ripped Open by an Iceberg

No one can tell me, nobody knows,
where the wind comes from, where the
wind goes.

Wind on the Hill, A.A. Milne

nlike the freshwater ice encountered in the mountains, sea ice is formed when the water temperature falls below −2.8°C (28.6°F) and in its first year, it has a greyish appearance and may become about 1.5 m (5 ft) thick. If it does not melt during the summer, it will become much denser in its second year and appear bluish. One reason is that sea ice initially contains pockets of brine that drain out after its first season. The brine pockets, though, provide sea ice with greater elasticity, allowing it to bend without breaking. For instance, 50 mm (2 in) of freshwater ice will support a person, while sea ice twice as thick will bend and the person's feet will sink into the water. When it fails,

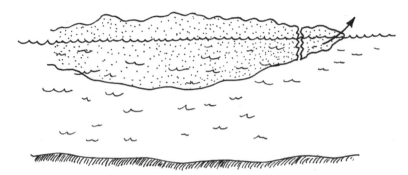

Fig. 9.1 Iceberg's child

freshwater ice does so suddenly in a brittle manner, while sea ice, with its brine pockets, cracks slowly.

In the Arctic, the polar ice sheet floats on the surface of the sea and drifts, seemingly randomly, driven by the currents. It is a frozen wasteland generally no more than 4.5 m (15 ft) thick but when driven by wind, snow and sea currents, it can raft up on itself, crack and pile into rip-rap hills as much as 15 m (50 ft) high. Under the weight of snow and piled-up broken ice, the great mass of the ice sheet is pushed down into the water and its thinner edges are bent upward. In the spring, as the ice sheet thins down, the bending strength of its outer edges may be exceeded, and chunks break off explosively, releasing a mass of ice that floats away (Fig. 9.1).

Just like the earth's tectonic plates, arctic ice is constantly on the move, sledding, cracking like thunder, opening invitingly before closing like a vise, but always offering areas of open water. For the unwary crew of a boat wandering into what looks like an open channel, the sudden closing action can crush a 250-ton ship in a few minutes. Some areas, called **polynyas**, remain open year-round, fed by unique currents and wind patterns. Experienced mariners learned to spot these clear areas by noting the abundance of sea birds and marine mammals that are instinctively drawn to them.

Farther south, but above the Arctic Circle in the North Atlantic Ocean, lies Greenland, the world's largest island, four fifths of which is covered by an ice sheet as deep as 3 300 m (11,000 ft). Pushed by the growth of the glacier-like ice, some sections of the ice descend into shoreline fjords as fast as 40 m (130 ft) per day and, in the summer,

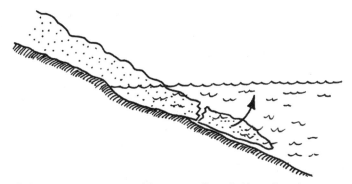

Fig. 9.2 Iceberg spawned from continental ice sheet

slide down into the Atlantic. As the ice descends into the deeper sea, ends are wrenched up by the buoyant force acting on the lighter ice, and chunks break off as the ice is unable to resist the resulting bending force (Fig. 9.2). From the north coast of Greenland and northwest coast of Ellesmere Island, ice islands as large as 770 km^2 (300 sq. mi) and as thick as 48 m (160 ft) thick break off the ice shelves that extend offshore. These may drift for decades before being carried southward in the East Greenland Current. From the western glaciers of Greenland, chunks of ice are calved into Disko and Melville bays and pushed northward by the West Greenland Current.

In the North Atlantic, these floating chunks of ice drift as **icebergs** and are driven by ocean currents, first northwestward along the Greenland coast, then southeastward along the North American coast. They continue on their journey following the Labrador current southward, skirting Newfoundland before entering the warmer waters of the Gulfstream that eat away at their evermore shrinking mass until they finally return to their watery state (as far south as latitude 40°N). This generally takes place in late spring and summer when the temperature in the Arctic rises, causing the Greenland glaciers to begin to move. The journey of the icebergs from spawning to meltdown may be as long as 3 200 km (2,000 mi) and take up to several months if the iceberg takes a direct route, or more than a year if the iceberg decides to drift around the west coast of Greenland for a while.

Icebergs are also spawned from the coast of the ice cap covering the Antarctic. With the weight of almost 2 km (1.2 mi) of ice pressing down upon it, most of the Antarctic is below sea level. Shelves

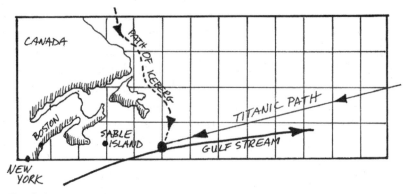

Fig. 9.3 Intersecting paths of the iceberg with the *Titanic*

of ice as much as 250 m (833 ft) thick overhang the coast so that
the ocean extends hundreds of kilometers under the shelf. The Ross
Ice Shelf, for instance, is the size of France and in October 1987
spawned an iceberg the size of New York's Long Island (155 km ×
35 km). The iceberg drifted northwestward for two years, crashing
into Victoria Land and breaking into three sections that were still
visible farther west the following year, having traveled almost 1 800
km (1,080 mi).

As we shall discuss later in this book, the extent of the Arctic ice
is rapidly shrinking, bringing with it dramatic changes to local climates. For instance, for 10,000 years, the Inuit language never had a
word for "robin," but now there are robins flying over Inuit villages.

Icebergs can have the appearance of a floating mountain or a
flat island, but show only ten percent of their mass above the water's
surface. This property makes them potentially dangerous because
an iceberg can easily extend horizontally far outside the edge that is
visible at the water line and can crush or rip the hull of a ship that
passes too close.

The Unsinkable Goliath

It was Benjamin Franklin who, in the eighteenth century, first
charted the existence of a strong current of warm water that provided Western Europe with a moderate climate, which was quite
exceptional for its northern latitude. The Gulf Stream skirts the
Grand Banks, south of Newfoundland, but like a slithering snake,
keeps changing its path so that at any given time its boundaries are
not clearly defined.

In the early spring of 1912, a number of icebergs were set afloat around Greenland and began their southward journey, riding the Labrador Current toward inevitable extinction in the warm waters of the Gulf Stream. At that time, it is quite likely that the path of the Gulf Stream was pushed southward by the cold Labrador Current, providing icebergs with an extended time of vigorous existence (Fig. 9.3). The speed of currents the icebergs were riding in their over-3 200 km (2,000 mi)-journey vary greatly in intensity with the seasons and the distance from shore. With speeds ranging from 0.1 to 1 m/sec (0.2 to 2 knots), we can assume that the icebergs headed for a historical rendezvous moved at an average of 50 km (30 mi) per day and had already been at sea for over two months since leaving their birthplace.

The *Titanic*, the pride of the White Star Line, was on her maiden voyage, having weighed anchor at Queenstown, Ireland at 1:30 the afternoon of April 11th on her way to New York with 2,206 passengers and crew. She was a huge ocean liner of 46,000 tons with three screws (propellers) pushing her at a full speed of 21 knots in calm seas. With four funnels towering 53 m ((175 ft) above her keel and a length of 269 m (883 ft), she was an imperial sight as she steamed across a calm sea. Approaching the area south of the Grand Banks, the wireless operator on the *Titanic* received a message from a nearby ship: "From *Mesaba* to *Titanic*. In Latitude 42°N to 41° 25', longitude 49°W to longitude 50° saw large icebergs, also field ice, weather good, clear." It was one of many messages warning of the presence of icebergs that were passed to Captain Edward J. Smith, who chose to continue pushing the ship along at full speed. At 10 PM on the evening of April 14th, the ship's lookout reported seeing a field of ice straight ahead. The officer on watch immediately ordered the engines stopped and then reversed and turned the ship full to port. The huge ship's inertia carried her forward while the slow turn prevented a head-on collision. However, the starboard side of the ship scraped along the ice with a slight grinding noise, but barely a tremor before the ship came to a stop. "It wasn't a terrible bump," related Margaret Moore, a survivor.

As the ship's engines stopped, there was an eerie silence. Most passengers and crew experienced the impact as a dull vibration coursing through the ship while in the boiler room it sounded like "the roar of thunder" as icy water sprayed through a series of slits that had been torn in the ship's side. We will never know for certain what actually happened, but based on observation of the sunken hull after it was discovered in 1985, it appears that the hull plates separated along riveted seams. Since the steel used at the time tended to become brittle at low temperature[19], it is likely that the force of the impact in the frigid waters caused the wrought-iron rivets to crack and pop out, wedging open the seams between the hull plates, allowing water to pour into the hull. Of the sixteen watertight compartments, the forward five and possibly six had been breached and the ship was fatally wounded. The designers had never considered that over four compartments could possibly be simultaneously breached. Recent exploration of the wreck has revealed that six open seams with a total length of about 30 m (100 ft) located along the joints between hull plates about 6 m (20 ft) below the water line allowed more than 39,000 tons of water to fill the hull, more than enough to sink her.

For two hours after striking the iceberg, the ship's band continued playing ragtime tunes as the passengers slowly awoke to the impending disaster. Finally, Captain Edward Smith ordered the lifeboats lowered, knowing full well that they had space for only half the passengers and crew. At about the same time, the wireless operator sent out an SOS, the first time in maritime history that such a signal had been used. At 1:55, the last lifeboat cleared the doomed hull. The ship began its final plunge at 2:18 AM, roaring and crunching into the icy North Atlantic. A survivor, Eva Hart, watched from a lifeboat and heard "the noise of drowning people.... There was so much screaming, crying and shouting." And then there was dreadful silence. After first breaking in two, the ship sank within two minutes and carried the remaining 1,522 passengers and crew to their grave, 3 810 m (12,500 ft) below the calm waters. The second officer, Charles Lightoller, witnessed the sinking: "Slowly and almost majestically, the immense stern reared itself up, with propellers and rudders clearing the water, till at last she assumed the exact perpendicular. Then with an ever-quickening glide, she slid beneath the water of the cold Atlantic.... Like a prayer as she disappeared, the words were breathed, 'she is gone'."

Only 705 survivors were rescued from the ship's twenty lifeboats by the Liner *Carpathia*. Had the ship an adequate number

---

19 See *Why Buildings Fall Down* for the story of the Silver Bridge that collapsed due to cold-weather embrittlement of steel.

of lifeboats, the two and a half hours that passed while the ship foundered in a smooth sea would have been enough to save all her passengers and crew. It is a cruel testimonial to the conceit of the designers and the operators who so ardently believed the ship to be unsinkable that they did not outfit it with sufficient lifeboats.

This momentous tragedy still evokes a unique mythical appeal among the many catastrophic events of the twentieth century. The search for the *Titanic's* remains became a passionate quest and seventy four years after her final plunge, she was discovered by a French-American scientific team, sadly silent as she lay partially buried in the muck at the bottom of the sea.

Strangely enough, the disaster of the *Titanic* was foretold in an 1898 book, *Futility*, written by Morgan Robertson, in which he described the story of an enormous ocean liner driven by three screws and with luxurious accommodations. The liner was reputed to be unsinkable because of her transverse waterproof bulkheads and therefore, like the *Titanic,* was equipped with only limited lifeboat capacity. With unique artistic premonition, the liner in Robertson's tale struck an iceberg on an April night and sank with half her passengers. Strangest of all was the name given by Robertson to his storybook ship; the *Titan.*

Icebergs Today

Following the path of the iceberg that caused the sinking of the *Titanic,* icebergs continue to descend from the shores of Greenland. Today, however, the ever-eroding edge of the ice cap spawns cordons of fresh icebergs, bringing huge amounts of fresh water into the North Atlantic. A hundred years ago the whole northern coast of Ellesmere Island was edged by a continuous ice shelf that is now almost 90% gone. This erosion of ice results in changes in the salinity of the water and in the global circulating pattern, the **thermohaline**[20] circulation (Fig. 14.8), part of which, the gulf stream, still brings warmth to northern Europe. This melting of ice is attributable to global warming and, as discussed later (Chapter 14), can have disastrous consequences.

Observing the melting taking place in Greenland, Dr. Joseph R. McConnell, a hydrologist, noted, "If Greenland melted, it'd raise sea levels by 20 feet. There goes Florida. There goes most of the

20   Thermo (heat), haline (salt)

Mississippi embayment. There go the islands of the South Pacific. Much of Bangladesh is obliterated. Manhattan would have to put up dikes." This is not far-fetched, as sea levels continue to rise about 2.8 mm (0.11 in) per year just due to thermal expansion. Above this is the added rise caused by the influx of melted ice with as-yet uncertain consequences. South Sea islanders are today planning the abandonment of their centuries-old homes as the rising ocean levels lick the shores of their lowland islands and, as this book is being written, some have already left.

10

Monsoon and Other Big Winds
Death and Renewal

Now I can no more close my eyes in slumber.
Now I know that monsoon showers of
    arrows
must batter my heart.

Rabindranath Tagore, *The Conch*

According to the generally accepted plate tectonic theory, about two hundred million years ago the Indian subcontinent broke free from the island of Madagascar in the Indian Ocean and began to move slowly toward the northeast. Upon reaching the Asian continent, about twenty million years ago, the northern edge of the subcontinent dipped down and pushed up the Tibetan Plateau (Qing Zang Gaoyuan) and masses of granite, which consequently broke up and folded, creating the great Himalayan mountain chain with its thirty peaks, almost nine kilometer highs, the highest on earth[21]. This plate movement, which is still continuing today, raises the mountain

---

21   See *Why the Earth Quakes*, Levy & Salvadori, WW Norton, 1995

chain ever higher.[22] The northern barrier created by the Himalayas, which extends in a roughly east-west direction, is the major reason for the unique climate conditions experienced by the Indian subcontinent. The mountains disrupt the northeast trade winds of the Hadley Cells, creating a barrier to the arctic winds that would otherwise flow down, as they do, for instance, across North America. Unable to benefit from these cooling winds, the Indian subcontinent's weather over the year ranges from hot to hotter.

Great rivers nourish the Indian subcontinent. In the north, the Indus River, fed by five rivers descending from the Tibetan Himalayas, flows 3 200 km (2,000 mi) in a southwesterly direction toward the Arabian Sea. Like the Nile, it annually overflowed and fertilized a broad valley; at least until present-day dams and storage basins tamed it. Also descending from the Himalayas, the sacred "mother" Ganga (Ganges) merges at Allahabad with her sister goddess Yamuna (Jumna) and, joined farther downstream by a third sister (Gomti), spills its waters among ten thousand rivulets in a broad delta in the Bay of Bengal. This is the river that carries the ashes of Hindu bodies from Benares, the holiest city in Hindu India. In the dry, desert-like central states, the central and southern rivers appear in the winter months as broad gravel ways with barely any water. But in the summer months, these rivers may suddenly become raging torrents, fed by the bounty of the monsoon rains (Fig. 10.1).

Monsoon

Ancient Arab traders plying the Indian Ocean observed that the prevailing winds in the summer months blew from the southwest, and in the winter months blew from the opposite direction. Since these winds defined the seasons, they named them *Mausim*, which is Arabic for season. Armed with this knowledge, they were able to efficiently schedule their voyages between India and Arabia and dominate the spice trade, selling their goods throughout the Mediterranean basin. In 40 AD, Hippalus, a Greek merchant, also discovered that the monsoons, which nourish India's pepper vines, reverse direction in midyear, something that the Arabs had long tried to obscure. Therefore, trips from Egypt's Red Sea coast to India and back could be shorter and safer than the Roman Em-

---

22   There are also sedimentary rocks containing fossils that have been found in the Himalayas. These may have been carried along during the uplifting process.

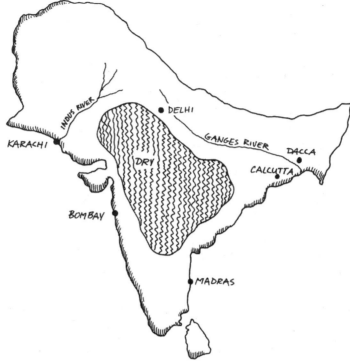

Fig. 10.1 Indian monsoons

pire had imagined. Taking advantage of this newfound knowledge, the Romans broke the Arab spice monopoly and started a booming trade with India.

Although it defines recurring winds, the word **monsoon** conjures up images of heavy rain, steaming as it strikes the starved, parched land, suddenly relieving it of the oppressive heat that had been building for months. It can, as well, represent death and destruction caused by either drought or flood if its natural cycle is upset. It is, in fact, one of the most dramatic climate phenomena on the planet, affecting 65% of the world's population. Although monsoons exist in North America and Africa, the most spectacular of all is the Asian monsoon, which affects a huge area that extends from Pakistan through India, Thailand, Vietnam, and even southeast China. Each country has even given the monsoon its own name: In China it is called *Mai-Yu*, in Japan, *Baiu*, and in Korea, *Changma*. With its contrasting wet summers and dry winters, it also plays a major role in modulating global weather. Weak rainfall and drought can result from warmer-than-normal temperatures in the Eastern Pacific, while cooler sea surface temperatures in that region may bring strong rainfall and floods. A high Eurasian snow cover has also been linked to a weak rainfall in the following summer's Indian monsoon. Finally, as we will see in Chapter 13, El Niño can also dramatically alter the Asian monsoon.

Summer Monsoon

Heat is an all-pervasive element of India's environment. Rajastan, a desert area south of Delhi and north of Ahmadabad, is a poor state, unable to support its populace. Nevertheless, its people display a high artistic spirit—peasants wear brilliant scarlet and saffron turbans and mirror-shimmering skirts and vests that contrast with the grey and mauve of nature's colors. It is a region that

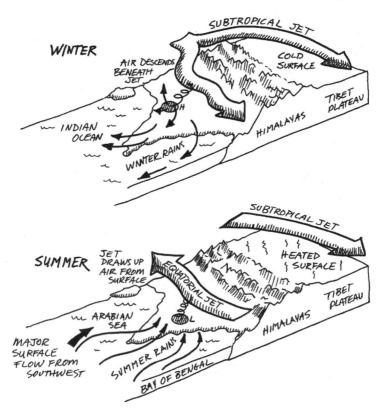

Fig. 10.2 Summer and Winter Monsoons

spawned bands of wandering musicians, dancers, and fortunetell-
ers, known today as Gypsies because they stopped for many years in
Egypt on the way to Rumania. The people of Rajastan bravely bear
the suffocating, hot, dry climate of India's winter while waiting for
a springtime change.

Beginning in March, the land of the Indian subcontinent is heated during the day by the subtropical sun. Although the adjacent sea
is also heated, the land will be heated more than the sea because the
heat absorbed by the sea is mixed and distributed through a greater
depth by convection. Consequently, over the next two months, the
land slowly absorbs much more radiant energy than the adjacent
ocean. This heated air rises above the land and is replaced by cooler
sea air flowing in from the southwest (Fig. 10.2). By the beginning of
June, the sea air, heavily laden with moisture, also rises as it meets

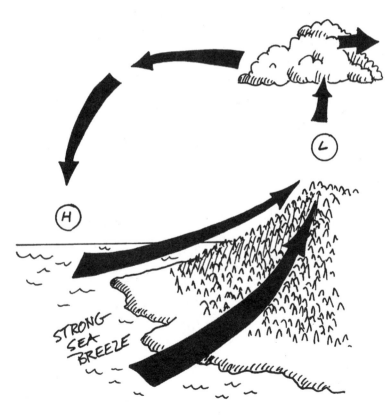

Fig. 10.3 Air-sea-land monsoon cycle

the lines of onshore mountains and cools as it ascends until it become fully saturated and condenses into rain (see p. 81) (Fig. 10.3).
Starting in Kerala on the southern tip of the Indian subcontinent,
grey-white monsoon clouds hang gloomily in the sky. In the valleys,
the clouds float just above the rivers and rice paddies, and a murky
dark film obscures the distant mountains. Throughout the land, the
air is heavy with hot dampness until one day a heavy shower brings
cool relief to an overheated populace and welcome nourishment to
parched crops. The rain-bearing monsoon winds then sweep up and
down the Indus River plain and the Ganges valley, where 75% of the
annual rain will fall in the following three months (Fig. 10.4). The
intensity of the almost-daily rainfall varies throughout the land.
The highest seasonal rainfall ever recorded was in Cherrapunji, in
the Dhasi hills, where an incredible 16 meters of rain fell in 1899.

Throughout the country, rice farmers are joyful, expecting a

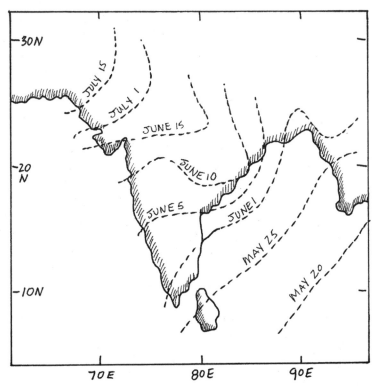

Fig. 10.4 Onset of Indian monsoon

good crop and praying that they have timed their planting correct-
ly: If the rains come too early, the seeds may not yet have been in the
furrows; if too late, the seeds may have been picked clean by crows.
They also hope that the rains will not result in torrential floods
that will damage or wash away the fragile plants. In the streets of
Mumbai, as the water rises, some people go without shoes as store
owners build small dams outside their shop doors to keep out the
rising water. Children in Mumbai will build paper boats from the
local newspaper and drop them in the puddles and rivulets, and
soon entire fleets travel around the city. In the wildlife sanctuaries
of the northern Indian state of Uttar Pradesh, tigers, wild boars and
monkeys flee from the rising waters seeking refuge in higher terrain. Torrential rainstorms may sometimes result in violent landslides and flooding, submerging thousands of villages, as happened
in 1998 along the Brahmaputra River. Along the coastline of the
region, fishing fleets may be forced back to port as the weather al-
ternates between hot sunny skies and wild, drenching rainstorms.

If the monsoon is late or unusually weak, drought may result.
If it is particularly heavy, devastating floods can occur. Such is the
power of the monsoon that in 1770, ten million people died of starvation in Bengal when the rains didn't come, and in 1943, three million perished in the famine caused by a light monsoon. The effects
of a light monsoon have recently been tempered with the construction of dams and storage basins, but flooding caused by an unusually heavy period of rain is still, occasionally, devastating. As recently
as 1998, thousands died and millions were left homeless as floods
destroyed farmland and ravaged the eastern state of Assam.

A heavy summer monsoon affects the whole Asian region, and,
apart from India and Bangladesh, has been particularly damaging to the regions adjacent to the two major river basins in China:
the Yangtze and the Yellow. With over five hundred million people
within their zone of influence, flooding along these rivers can be
disastrous, as it was in 1998 with over 4,000 lives lost and billions of
dollars in property damage.

Winter Monsoon

By October, the rains have stopped and, as fall gives way to winter, the land cools and the winds begin to flow out toward the sea,
auguring the arrival of the northeast monsoon. Until May, cooler,

dry weather will cover the land, which will then be warmed by the sun as the cycle repeats once more.

Other Winds

Cyclical winds called monsoons also exist outside the Asian continent. Northern Australia, West Africa, parts of East Africa, and the Gulf of Mexico all experience monsoon conditions. None is as well developed as the Asian monsoon but each has a distinctive characteristic. In the sixteenth century, explorers used the monsoon to round the Cape of Good Hope in an eastwardly direction after having explored the west coast of Africa (p. 25–26). Alexander the Great's admiral discovered how to use the monsoon by following the coast from Arabia to India. Fourth-century ships sailed on monsoon winds from Malobia in East Africa to China. In the United States, the Arizona monsoon, in the summer, looks like a river of cumulonimbus clouds that first runs up along the spine of Mexico and then spreads across New Mexico and Arizona, triggering monsoonal storms.

The monsoon, though, is not the only recurrent wind on the planet:

- In the eastern Mediterranean, the **Etesian** (from the Greek, *etos* or season) winds blow in the summer from the north over the hot Sahara.

- In the early spring, the **Sirocco**, a hot, dry wind, blows north from the Sahara and Arabian Desert across the Mediterranean Sea. When Michelangelo was working on the frescoes of the Sistine Chapel, the Sirocco blew sea salt onto the surface, causing disintegration and flaking, delaying the completion of the work. Another dramatic example occurred in 1901, with a Sirocco that dropped two million tons of dust and sand from the Sahara across Europe.

- The **Mistral** (from the Latin, *magistralis* or masterly) blows southward from the colder regions of northern Europe across the plateau of central France through the Rhône Valley to its delta and Provence. It blows in the spring and fall with gale-force winds (Beaufort 10). The Mistral is celebrated in the literature and art of Provence by such artists as Vincent Van Gogh and writers such as Frederic Mistral. In Italy, the winter wind that blows south from the Alps is called the *tramontana*, and brings frigid temperatures to the normally temperate regions.

⇛ **Foehn** winds drive down from mountain slopes and gain heat by compression. Forced upward on the windward slope, moist air rises and drops moisture as it moves from a high to a low-pressure area over a mountain and in the process, is cooled at about $0.6°C/100m$. On the leeward side of the mountain, it is then compressed and heated adiabatically (at $1°C/100m$, $5.5°F/1000ft$). The result is a hot dry wind blowing at the base of the mountains. For instance, at the base of the Alps, velocities of foehn winds have been known to reach hurricane levels.

⇛ Foehn winds also exist in the Rocky Mountains of Montana, Colorado, Wyoming, and Canada, where they are called **Chinook** winds. Some of these foehn winds are associated with **leewaves** that bring air down from higher in the atmosphere. As these winds descend, they are compressed and heated, becoming quite warm and strong as they reach the surface.

⇛ In Austria and Switzerland, such winds are called **Schneefresser** (snow eater) because they quickly melt snow, causing mountain streams to flow as torrents.

⇛ In the Andes Mountains of Argentina and Chile, foehn winds are known as **Zonda**.

⇛ The **Santa Ana** winds on the western slopes of Southern California are another type of foehn wind. These blow from the north or northeast, sometimes at gale velocities, drying all vegetation and sucking up moisture from the ground. This results in blowing, choking dust and stinging sand. During the winter, Santa Ana winds blow from the hot deserts and plateaus in the north and descend through mountain passes (Cajon and Santa Ana) becoming hotter as a result of compression. The drying effect of Santa Ana winds often result in dangerous fire conditions. In 1961, such a wind blew through BelAir and Brentwood fostering a fire that destroyed 450 houses and stopped only when the winds died down. The recent period of drought that started in 1999 has had a devastating effect with yearly fires causing ever greater damage. Whether this drought is attributable to global warming, or is a cyclical event that has been documented in the tree ring history of the region, is not clear.[23]

⇒ The **Loo** is a hot dusty wind that blows in the region of Bihar, India as the sun climbs northward.

Of all these winds, none more clearly exemplifies the coupling of ocean-atmosphere-land systems nor has more dramatic global effect than the monsoon.

It should be noted, though, that flooding resulting from monsoons is by no means unique, as storms sweeping over the globe can cause floods of equal or greater power to those attributable to monsoons.

---

23   Dendrochronologists (who study tree rings) have determined that drought is more prevalent than wet periods and in the Southwestern United States region occur, on average, twice each century. Also, droughts lasting over a century have been recorded a number of times in the last thousand years.

11

The Cradle of Civilization
Floods Enrich the Nile Valley

Very many times the Nile
and other very large rivers
have poured out their
whole element of water
and restored it to the sea.

Leonardo da Vinci

Fueled by the sun, water is the medium that circulates between earth and sky. The earth's water cycle begins as evaporation that lifts water vapor into the atmosphere mainly from oceans but also from lakes and the land itself. Carried by wind and atmospheric currents, clouds into which this vapor has condensed circle the globe. Drop by drop, this water eventually falls onto the earth as rain or snow, and as it rains down onto lakes and rivers it also seeps into the earth. Once the earth is saturated, water flows on the surface in rivulets that merge into streams and creeks. Pulled downhill by the force of gravity, these streams descend from hills and mountains and gather into rivers. In their inexorable drive toward the ocean, the largest and strongest of these rivers collect many smaller rivers into their fold, becoming giants that drain large parts of the landmass of our world. (The Mississippi River, for instance, drains about 40% of the Continental United States.) These great rivers—the Amazon, Congo, Mississippi, Nile, Yangtze, Ganges and Mekong—both nourish the land and refresh the oceans of the world, completing the earth's water cycle (Fig. 11.1).

River	Length Km (mi)	Drainage Basin Square km (square miles)
Amazon	6400 (4,000)	6968000 (2,722,000)
Congo	4800 (3,000)	3648000 (1,425,000)
Mississippi-Missouri	6720 (4,200)	3174000 (1,240,000)
Rio de la Plata	3680 (2,300)	3067000 (1,198,000)
Ob	3840 (2,400)	2944000 (1,150,000)
Nile	5600 (3,500)	2834000 (1,107,000)

The World's Largest Rivers

Always drawn down by gravity, a mountain stream cuts a path through steep terrain and rushes downstream, cleaning its bed of earth particles, leaving only rocks along its path. Where the land is steep, there is a rapid, with water churning around rocks. Where the land suddenly falls away, a waterfall—such as the majestic Niagara or Victoria—leaps downward, partially obscured behind a fine, translucent mist. The faster the river flows, the more earth particles it carries (including rocks) that it will deposit in the bed of its shallower and slower-moving sections, and always in its delta where the river flows into the ocean. The speed of a section of river can be measured by noting the deposited material: rocks and pebbles in the riverbed of the rapids, sand along the intermediate sections, and silt and mud on the bed of the slowest-flowing sections. In the delta region, since the ocean level changes with the tides, a reverse flow upriver takes place at high tide (flood tide) and the river's flow speeds up at low tide (ebb tide). These flood and ebb tides near the river's outlet have a scouring effect on the riverbed, stirring up the deposited material and constantly changing the topography of the delta with its many-fingered channels.

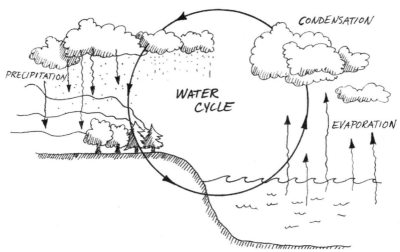

Fig. 11.1 The earth's water cycle

The Nile

Of all the world's rivers, the Nile has the longest history of human interaction. Homer wrote of "Egypt's heaven descended spring." Located at the heart of the earliest civilization in the Mediterranean basin, the river nurtured the growth of human settlement. It provided water in an otherwise arid landscape to meet the needs of early Egyptians, enriching the land, permitting farming to flourish, and attracting animals that were exploited as a food source as well as beasts of burden. There were camels, donkeys, hippos, giraffes, scorpions, birds, buffalo, and the dangerous crocodiles that were both feared and worshiped: The crocodiles even ate children who swam too close. Without the Nile providing the only water in the region, there would not have been an ancient Egyptian civilization and the history of the world would have been very different.

The delta region of the Nile is influenced by its proximity to the Mediterranean Sea and has a winter rainy season, but the upper Nile is generally dry. In ancient times, the whole region was covered by a primeval forest that was, over time, cleared and may have contributed to a climatic change and a reduction in rainfall. Today's arid climate is the result. The Nile valley is confined in the west by the Sahara Desert and in the east by the Arabian Desert, with sand constantly trying to infiltrate from both sides.

In Upper Egypt, above Aswan, the river cuts through Nubian sandstone. Downstream, the river valley is framed by limestone cliffs that were used by the ancient Egyptians both as a source of stone and as the site for tombs—which were cut into the cliffs.

Within this seemingly hostile environment, the river flows, delivering water from the tropical central African highlands to the Mediterranean Sea. Summer rains wash down from the hills into the two branches of the river: the Blue Nile descending from Lake Tana in Ethiopia and the White Nile originating in Burundi, south of Lake Victoria. As the tropical rains in central Africa swell its two branches, the river speeds past the confluence at Khartoum, Sudan, and its waters rush through cataracts before reaching the broad valley below Aswan. In the middle of the summer, unable to contain the rising waters, the Nile used to gently overflow its banks, spreading its black muddy waters over the surrounding lowlands. These annual floods deposited sediments that contributed to the valley's fertile black soil. Construction of the Aswan high dam, which was completed in 1971 with its containment reservoir sufficient to store a season's floodwaters, has dramatically altered the flow of the river. No longer do the soils and nutrients descend to the fertile valley but remain trapped behind the dam, unable to fulfill their destiny. Over time this silting will undoubtedly reduce the size of the reservoir and the effectiveness of the dam. Another solution will then be needed to control the annual floods of the Nile.

Meanwhile, the waters of the Nile are still used to irrigate the land as they have been since at least 1300 BC. Instead of relying on the uncontrollable floods to irrigate the land, water is now stored in basins and canals for later use. Yet the nutrients that used to feed the fertile valley are now trapped behind Aswan, so man-made fertilizers must now be added to the earth, replacing what nature customarily provided.

From December to March, the air over the upper Nile is cool, but during the rest of the year the temperature is very hot. The heat is somewhat alleviated by strong southeast winds that blow from the middle of February to the middle of June, rising to almost hurricane intensity and filling the air with dust. The balance of the year there is a northwest wind that mitigates the intense heat of the day. The ancient Egyptians valued, as one of the best things in life, to "breathe its sweet breath." The annual floods, which would reach their high point in mid September and last until the end of the year, spread coolness, dampness and fertility throughout the valley. It is not surprising, then, that the ancient Egyptians dated their New Year from the time of the highest flood. It was such an important event that *Nilometers,* under the protection of the State, were installed to measure the level of the flood. At Aswan, one such gauge consisted of a staircase hewn out of the rock on which the dates of ancient floods are inscribed.

Dr. Livingtone, I Presume

The search for the headwaters of the Nile engaged ancient Egyptians and became the Holy Grail of the nineteenth century. In the middle of the second century, a Greco-Egyptian mathematician/geographer, Ptolemy, summarized all that was known and surmised about the trace of the Nile. He wrote that the White Nile originated in two lakes in the equatorial region of Africa that were fed by streams descending from the snow-capped Mountains of the Moon. He further wrote that the river, emanating from those lakes, joined together with another river (the Blue Nile) originating from a lake on the equator to become the lower Nile. It was an amazingly accurate picture, considering the scarcity of information that was available at the time. For the next fourteen centuries, nothing of note was added to complete and clarify Ptolemy's vision.

The source of the Blue Nile, or Abay Wenz (meaning "great river"), as it is known in Ethiopia, was long known to spring from the face of Mount Gishe, above Lake Tana. On the calm lake, natives still paddle around on *tankwas,* or papyrus boats, as their ancestors did thousands of years earlier. Below the southwest corner of the lake, the river drops over the Tis Isat Falls and snakes its way in a wide arc following a 900-km (560-mile) journey through the Ethiopian Highlands. Violently churning its way through canyons of black volcanic rock on its way to Khartoum, the river drops 1 200 m (4,000 ft) and, laden with silt, appears more like an agitated brown soup than the docile Blue Nile.

The first intensive efforts to discover the source of the White Nile were undertaken in the late eighteenth century, first by the French, followed by the British and the Egyptians. After many unsuccessful attempts to discover the source by following the river southward from its delta, Richard Burton and John Speke, in 1857, started instead from Zanzibar, off the east coast of Africa. They assembled a caravan and traveled west from village to village until reaching Tabora, where they were told that there were great lakes west and north of their route. Burton, as the leader of the expedition, chose to go west and reached Lake Tanganyika. The explorers sailed to the northern end of the lake looking for an outlet that could be identified as the beginning of the Nile. It wasn't to be found. By now, both men were sick, and Burton decided to stay and compile his notes describing the discovery of Lake Tanganyika. After some weeks, Speke decided to investigate the reports that another lake existed farther north. After three weeks he reached a large Nyanza (lake) that he rightly observed to be the main reservoir for the Nile River, and which he named Victoria Nyanza, in honor of his queen. He hadn't spent enough time to properly explore it or even to look for the outlet, but he felt certain that he was right. When he rejoined Burton, the two argued about whose lake was the true source of the Nile. Burton admitted that maybe Speke's lake was a source but that maybe there was a river between the two lakes and therefore Tanganyika was also a reservoir. When Speke returned to England, he convinced the Royal Geographical Society to mount another expedition that, this time, he would lead. After a harrowing two-year trek in 1862 he properly explored "his" lake and found the Nile outlet by Ripon Falls. To be certain, he followed the river to Gondokoro, and returned triumphant to England. Burton still challenged Speke's conclusions, claiming that he had traveled overland to Gondokoro and only occasionally spotted the river. Perhaps there was more than one river and even possibly more than one lake. To settle this disagreement, a debate was proposed before the British Association for the Advancement of Science at Bath. An audience of scientists and geographers would evaluate the arguments presented by these two gentlemen, and would decide who is more correct. The scene was set for what was called the "Nile Duel." Tragically, on the day before the September meeting, Speke died in a hunting accident when he fell on his rifle while climbing over a wall. For a long time thereafter, some people thought that he had committed suicide rather than face the scrutiny of a scientific debate.

The meeting on the date of the now-cancelled debate was restrained, and among the papers presented was one by a missionary doctor, David Livingstone, a firm abolitionist who read a paper on the evils of the slave trade. The 52-year-old Livingstone had spent twelve years as an explorer, tracing the Zambezi River and discovering Lake Ngami in southern Africa. He had a neutral position on the source-of-the-Nile controversy. Therefore, Sir Roderick Murchison, the head of the Royal Geographical Society, asked Livingstone to undertake an expedition to settle the dispute. Livingstone accepted, in the firm belief that he could thus help abolish the Arab slave trade, and that it was in effect a holy undertaking to find the source of "Egypt's heaven descendent spring." On January 28, 1866, he arrived in Zanzibar, the capital of the Arab slave trade, and assembled a modest sixty-person caravan. In spite of Livingstone's views on the slave trade, the Arabs helped him to assemble and equip the party.

Livingstone was predisposed to the view that the source of the Nile was very far south of Lake Victoria, and therefore headed west to the southern tip of Lake Tanganyika. For nine months he traveled about 1,300 km (800 miles), suffering all the while from dysentery and malaria. Reaching the Chambezi River, he followed one river after another looking for one that flowed north into Lake Tanganyika. Two years into his trip, he gathered intelligence that led him to believe that the Nile's source was not as far south of Lake Victoria as he originally assumed and realized that he had been wandering aimlessly. During the next two years although he suffered incredible agonies, he doggedly plodded ahead searching for a mythical river that flowed north: He was constantly ill; he had to walk while suffering with ulcerated feet; he lost his medical supplies and chronometer when some of his porters deserted and ran away; he was threatened by cannibals and other local tribes brandishing poison spears and arrows; he faced starvation when he was unable to restock supplies that had been taken by marauding tribesmen; he also had to improvise constantly as he had to negotiate with the local tribes to pass through their territory, and for canoes and rafts to cross rivers and lakes.

Sick and exhausted and "reduced to a skeleton," he arrived in Ujiji on the eastern shore of Lake Tanganyika in the fall of 1871 with Chuma and Susi, the last two members of his expedition (Fig. 11.2). Livingston had been in the African interior for five years with only occasional rumors and speculation of his whereabouts and his health filtering back to London. Sensing that there was a story in this tale of the wandering explorer, the *New York Herald* decided to sponsor an expedition to find the good doctor, and assigned the task to its star reporter, Henry Morton Stanley.

Stanley wondered, "...at the cool order of sending one to Central Africa to search for a man whom I, in common with almost all other men, believed to be dead." Given all the money he needed for such an expedition, Stanley, who had no experience as an explorer, accepted the challenge. He assembled a party of almost 200 porters and set out from Zanzibar in January of 1871 on what was to be only the first of his many African adventures. Traveling for almost nine months, he arrived in Ujiji, by a miraculous coincidence, less than a week after Livingstone. Had he arrived much earlier, he may never have encountered Livingstone and reported his greeting, "Dr. Livingstone, I presume."

When Stanley departed five months later to report his findings, Livingstone stayed behind, never again left Africa, and died a year later.

The spirit of the adventure surrounding the exploration of the source of the Nile excited the popular imagination as a result of the stories published by Stanley. But there still remained questions about the Nile's source that could be resolved only with further exploration. Stanley was now eager to undertake the task and was able to raise the considerable funds needed. In the course of the next several years, he not only confirmed that Lake Victoria was the principal reservoir for the Nile, but also followed the course of the Congo River from its source to the sea. The ace reporter had become the most famous African explorer of the nineteenth century.

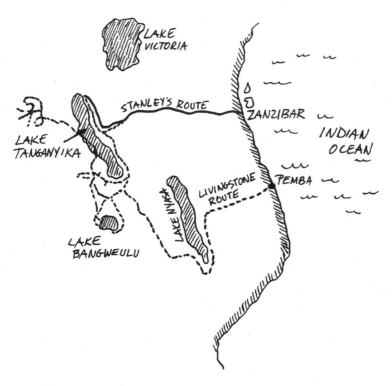

Fig. 11.2 Map based on contemporary information showing the route taken by Livingstone, 1866–1873, and the route taken by Stanley, 1871.

12

The Modern Floods
The Missouri and the Mississippi

> If by your art, my dearest father, you have
> Put the wild waters in this roar, allay them.
>
> *The Tempest*, Shakespeare

Floods cause nine times more deaths than either earthquakes or hurricanes. On June 15, 1991, Mt. Pinatubo in the Philippines erupted after a six-hundred-year sleep.[24] There were few casualties, thanks to the advance warning provided by the volcano's rumbling and puffing. Yet in that same year, 150,000 people died as a consequence of flooding that followed a cyclone in Bangladesh. It was small comfort that twenty-one years earlier, a similar flood cause twice as many casualties. No part of the globe is immune from the effects of floods.

---

24   See, *Why the Earth Quakes* by Levy & Salvadori for the story of earthquakes and volcanoes.

A 1952 flood in Lynmouth, England, caused 34 deaths when three times as much water as the Thames carries in three months raged through the town.

In 1976, a 250 mm (10 in) rainfall caused 139 deaths as the Big Thompson River in Colorado emulsified vehicles and carried away over four hundred homes. The wind, in this case, came in from the east, bringing moisture from the Gulf of Mexico. As the land rises when going westward in Colorado, this moist air condensed into rain. The sky turned a washed-out blue and the storm remained stationary over the mountains, dropping rain west of Loveland.

In the Italian Piedmont in 1994, 625 mm (25 in) of rain fell in a sixty-hour period, more water than the 1966 deluge of Florence that killed 113 people and ruined priceless works of art. This time, more than sixty people died and the flood left a devastated landscape that was as damaging to the area as World War II.

China has long battled monsoon-fed periodic floods of its two great rivers, the Yellow and the Yangtze. In 1931, when the Yangtze and Huai Rivers roared out of their banks, 225,000 people died and many more were left homeless. The intensity of the flooding of the Yangtze River is the result of two natural phenomena. First, the Tibetan-Quinghai Plateau, which is the source of the river, lies 6000 m (20,000 ft) above the heavily populated alluvial plain and causes a rushing flow of water to occur during the monsoon season. Second, the effect of El Niño (see Chapter 13) has intensified the monsoons that drench central and southern China. These effects were exacerbated by a man-made event: Forest-cutting in the upper reaches of the Yangtze has intensified its flow because of increased runoff.

Over the past two thousand years, northern and central China have suffered 214 large floods. To change this historical inevitability, in 1910, Dr. Sun Yat-sen, the founder of modern China, proposed the idea of a water-control project along the Yangtze. Interrupted by wars and revolutions, it would be another eighty-four years before such a project would begin.

Three Gorges Dam

A dam, 185 meters high, has now been constructed to contain a reservoir of over one thousand square kilometers that will house the world's largest hydroelectric generation plant, with a capacity of 18,000 megawatts. The dam's location will ease the risk of flooding for the middle reaches of the river, but has destroyed the Three Gorges, a 220-km stretch of unparalleled beauty celebrated in many Chinese paintings, with an untold wealth of historic, cultural and archeological treasures. The dam also required the resettlement of over a million people displaced by rising waters of the 600-km-long reservoir. It is the most ecologically destructive project since the construction of the Aswan high dam that flooded a significant portion of Egypt's historical region. Nevertheless, the Three Gorges dam will certainly alleviate the Yangtze's periodic floods and will provide a significant increase in the country's greenhouse-gas-free power supply.

Big Muddy

The Mississippi River drains an incredible 40% of the land area of the United States, a drainage area twice that of the river Nile in Egypt or the Ganges in India and larger than the Yellow River in China. Only the Amazon and Congo Rivers can claim to encompass a larger drainage basin, and those rivers flow mostly through a thinly occupied landscape, while the banks of the Mississippi and its tributaries, the Missouri and Ohio rivers, are crowded with cities, towns, and millions of people. From the headwaters of the Missouri river in the Rocky Mountains, the combined Missouri-Mississippi River stretches 6020 km (3,740 mi) to the delta in the Gulf of Mexico (Fig. 12.1).

In 1541, Hernando de Soto, a dashing young conquistador, led an expedition that landed in Florida in search of gold and other treasures. Over the next two years, the party moved north and then ventured west from Alabama until reaching the barrier of the Mississippi River. Unable to cross the mighty river in mid-March, one of the expedition's members wrote, "Then God, our Lord, hindered the work with a mighty flood of the great river, which...came down with an enormous increase of water, which in the beginning overflowed the wide level ground between the river and the cliffs." The river then overtopped its banks and inundated the town of Aminoya, an Indian village, and spread 100 km (60 mi) across the broad, flat land. The chronicler was also told by an old Indian woman that such a flood was experienced every fourteen years. (In reality, the periodic flooding of the river is typically caused by spring rains and

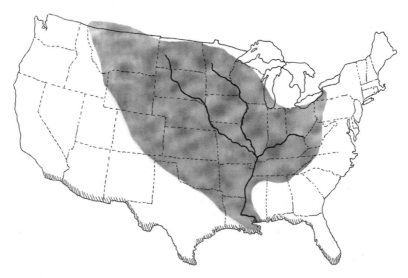

Fig. 12.1 The Mississippi River Zone of Drainage

melting snow, and does not occur precisely every fourteen years.) Since first occupying the two broad plains on the sides of the river, the early European settlers tried to control the periodic flooding by building dams and erecting levees along its banks. But the river kept challenging the builders to erect higher and higher levees and more dams as the waters rose to new heights year after year. In the late summer of 1926 a chain of meteorological events began that led to the greatest flood yet experienced in the region. In Hindu mythology such an apocalyptic flood would signify the end of the world—and so it seemed to those who lived through it.

As summer waned in 1926, a low-pressure system covered the central United States and caused heavy rain to fall over a large region from Nebraska and South Dakota moving eastward to Ohio. The rain came down in sheets and was accompanied by booming thunderstorms. The storm lasted for days and was relieved for only forty-eight hours as a second low-pressure system moved up the Mississippi valley. For the last two weeks of August, the rain soaked into the soil until it was completely saturated and water began flowing across the land into rivers and streams that rose and spread beyond their banks. One storm followed another without relief through early September, flooding large regions of Nebraska,

Kansas, Iowa, Illinois and Indiana. At first, the Mississippi River was able to carry the accumulated rainwater but slowly, it began to bloat and fill its channel. Then, in October, the section above the confluence of the Ohio and Mississippi River overtopped its banks and spread water over land that was already completely saturated. In the lower Mississippi (below the juncture with the Ohio River), levees as high as a three-story building contained the flow and were thought to be impregnable. Throughout that fall, the river remained high and reached levels not usually attained until the time of the spring thaw and rains. Yet, as the rains ceased in late October, re-lieved officials announced that there was no cause for alarm. But it was only a temporary pause, as 1926 was an El Niño year with its characteristic of persistence: If the rains start, they will continue!

As fall turned to winter, a deep cold front caused a rapid drop in temperature over South Dakota that was followed by a heavy snow-storm with 750 mm (30 in) falling on Helena, Montana. At the same time, torrential rains drenched the warmer parts of the Midwest. The streams and rivers throughout the Mississippi River drainage area could no longer contain the quantity of water flowing into them and began to overflow: The Big Sandy, the Cumberland, the Tennes-see, and Yazoo Rivers all overtopped their banks. As new storms raged in the New Year, more rivers flooded: The Allegheny, Monon-gahela, Illinois, Ohio, White and Little Red rivers all reached flood stage. As all these rivers poured their waters into the Mississippi, it, too, reached flood stage and remained so for a record 153 days.

Severe weather continued unabated for the next three months. Heavy snows blanketed the northern states while rain persisted in the South. Throughout March and early April it poured almost daily. "After a very stormy day yesterday," recorded Henry W. Ball in his diary on March 12, "it began to pour in torrents about sunset and rained very hearty until 10…. At daylight, a steady unrelenting flood came down for four hours. I don't believe I ever saw so much rain." Ball's diary monotonously repeats the words: rain, showers, pouring, torrents, on almost every day during that month and into April. On April 8th, he noted, "At 12 it commenced to rain hard. I have seldom seen a more incessant and heavy downpour until the present moment. I have observed that the river is high and it is al-ways raining…we have heavy showers and torrential downpours al-most every day and night…the water is now at the top of the levee."

Ball lived in Greenville, halfway between Cairo, at the confluence of the Ohio River, and New Orleans.

Finally, on April 15th, another storm poured 150–300 mm (6–12 in) of rain over several hundred thousand square kilometers—from Illinois to the Gulf of Mexico and from West Texas to Alabama.

Fifty-six hundred km (3,500 mi) of levees protected the land on either side of the Mississippi and its tributaries. These man-made embankments of soil, clay and sand had been built to withstand a fifty-year, but not such a cataclysmic, flood. Another issue threatened the levees: The river had many bends, some in the shape of a horseshoe. The faster the river flowed, the more the banks of the levees along the bends were eroded. Unable to withstand the incredible force of the river, levees in Arkansas were breached. All along the length of the river, heavy earth-moving machinery piled mountains of earth on top of the threatened levees and armies of workers filled sandbags and stacked them on the levees to stave off the advancing flood crest. But the rains did not cease, and on April 16th, a section of the levee below Cairo was ruptured, flooding the adjacent plain. This signaled the start of a catastrophe. In the succeeding days other sections of levee would fall to the incredible power of the swollen river that now carried two-to-three times the volume of water it can accommodate. Water roared through a crevasse in the breached levee with such force that it pulled debris, trees, and boats that ventured too close, and disgorged them onto the plain beyond. The crevasse just below the Mississippi's confluence with the Arkansas River was so large that it allowed the river to flood over a million hectares with six meters of water. As the rivers flooded, the center of the country was becoming an immense lake the size of Massachusetts, Connecticut, New Hampshire, and Vermont combined (Fig. 12.2).

Over 300,000 people had to be rescued from rooftops and over 300 died in the flood. To save the city of New Orleans from flooding, a levee on the east bank of the river protecting the adjacent townships of St. Bernard and Plaquemines was dynamited. In effect, these townships were sacrificed in favor of relieving pressure on the levees surrounding the more politically powerful city of New Orleans. This proved to have caused needless destruction, as the force of the river, naturally relieving the pressure on the New Orleans levees, soon breached levees north of the city.

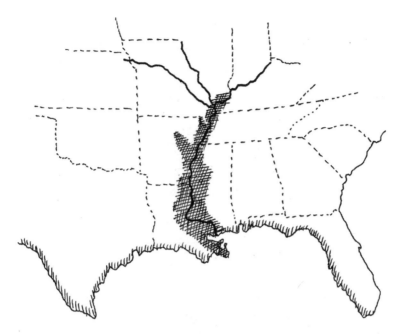

Fig. 12.2 The 1927 Mississippi Flood

By this time, the disastrous extent of the flood became apparent and President Coolidge assigned the task of coordinating relief
efforts to his secretary of Commerce, Herbert Hoover. A mining
engineer by training and a devout Quaker, Hoover had, throughout
his life, engaged in a number of humanitarian efforts combining
his rational approach with a moral conscience. At the same time
he was a supremely ambitious man and saw in the task before him
the opportunity for national prominence. In the next few months,
he would use public relations, expanding his role in the relief effort,
to boost his chances for the next presidential nomination. He organized the resources of the Red Cross, the military, the railroads, the
radio stations, and the National Guard for a massive relief effort,
and to feed and house the tens of thousands of refugees. All the
while, the river kept up its assault, overtopping a levee here, breach-
ing a levee there. To contain the assault, thousands of men continued to shore up the weakened levees. As April turned to May, more
rain fell with almost 300 mm (12 in) falling in a two-day period.
This resulted in miles of levees crumbling, causing the evacuation

of over a hundred thousand people. Only as the rains abated in the month of May was there any relief. The water, lying over fully saturated soil, would, however, continue to flood the land until August. Throughout that time, Hoover's name was constantly in the news, a fact that undoubtedly contributed to his eventual nomination for the presidency of the United States.

In the year following the flood, new measures were instituted to control the river in the future: Cutoffs, eliminating many of the sharp curves, shortened the river by 240 km (150 miles); levees were strengthened and increased in height; "fuses" were built into the levees by deliberately lowering sections that were expected to fail in a heavy flood; and diversion structures were built to divert flow into adjacent lakes and rivers.

In 1993, an El Niño event tested the effectiveness of these flood control measures. One of these, a spillway that diverts floodwaters above New Orleans into Lake Pontchartrain, proved totally effective. Others proved not to be as effective as the river, once again, demonstrated its power, causing disastrous floods in the whole basin. It was neither the first nor the last Mississippi flood since 1927, but to those who had experienced the events of 1927 this new flood seemed to be but a modest aftershock.

Many of these same flood-control measures failed in 2005 as Hurricane Katrina attacked New Orleans from the ocean rather than from the river (see p. 155).

13

El Niño
The Mysterious Current

> The wild child brings
> Chaos to the world.
>
> Anon

As we approach the issue of global warming, we are reminded that the earth is ONE and what happens in one corner of our planet often affects what happens halfway around the globe. This is abundantly clear as we learned that storms that originate in Africa can become deadly hurricanes striking the North American continent, and that humid winds originating in the Arabian Sea can become the monsoon that nourishes the Indian subcontinent. But, the mysterious current that is known as El Niño comes closest to defining our interdependent world.

Something's Wrong

Early in the 1983 New Year, fishermen returning to the port of Paita on the Pacific Coast of Peru appeared dejected. As they steamed into port, the black waterline painted on the side of the boat's hull was clearly visible and well above the surface of the water. This was because the holds of their ships contained barely one third of the normal catch of anchovies. Although located just south of the equator, the ocean waters off the Peruvian coast are usually cool and rich in nutrients that provide a bountiful harvest of fish. The reason for this apparent anomaly can be explained by understanding the immense coupled air-ocean system that exists in the pacific equatorial belt. Driven by the trade winds, the surface waters in the tropical Pacific are normally pushed along and warmed by the sun on their westward journey. The upper ocean layer, which is exposed to the warming effect of the sun, is separated from the deeper, colder region by the **thermocline.** As the trade winds drive warm water westward, the thermocline rises in the east, resulting in cooler near-surface waters. At this time, the warm upper layer is as much as 150 m (500 ft) deep in the west but as shallow as 30 m (100 ft) deep in the east. In effect, the winds determine the water temperature but, as we will discover, the water temperature also determines the winds. Along with the trade winds from the east, Coriolis accelerations cause upwelling along the equator. As the east winds push the surface water westward, the water turns right (northward) in the Northern Hemisphere, and left (southward) in the Southern Hemisphere. The resulting divergence at the surface brings cooler water coming up from below. (This can be seen from satellite images both in terms of cooler sea-surface temperatures and in terms of more living organisms in the surface water, seen as nutrient-enriched water). This creates a westward extended appendage to the cool water, brought upward along the western edge of South America.

Moreover, because of the pressure exerted by the trade winds, the ocean surface around Indonesia and the other islands of the western Pacific may actually be a half meter higher and 8°C warmer than the ocean off South America (Fig. 13.1).

Another consequence of the coupled air-ocean system is that humid air, fueled by the warm waters of the western Pacific, rises and forms the clouds that result in the rainfall and monsoon weath-

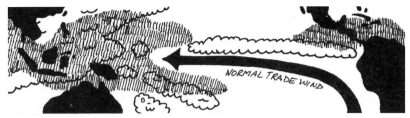

Fig. 13.1 Normal trade winds

er that is characteristic of the Asian subcontinent. Meanwhile, a reverse convective current of deep cool water usually flows along the ocean floor and rises up along the coast of South America. This upwelling of cool water completes the "normal" air-ocean circulation in the equatorial Pacific Ocean and results in the generally dry atmosphere usually experienced along the coasts of Ecuador, southern Columbia, and northern Peru as well as accommodating the abundant coastal fish population.

However, 1983 was not a normal year.

The southeast trade winds, in late 1982, were weaker than normal, allowing the tongue of cool water that usually extends westward from the South American coast to recede eastward. As a result, the thermocline lowered, moving a 150-m-(450-ft)-thick slab of warm water toward the coast of Ecuador and Peru, causing the sea temperatures to rise (Fig. 13.2) and the fish harvest to decrease.

The weaker trade winds also allowed the piled-up warm Pacific waters to slosh to the east and let the hot-air-fueled storm engine that feeds the monsoon to move eastward as well. These two effects totally changed weather patterns in the Western Pacific. Because the Pacific Ocean is huge, and therefore the power represented by the heat content of its mass of water is so enormous, a change or

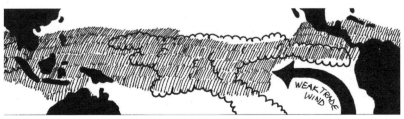

Fig. 13.2 Weak trade winds in an El Niño year

disruption of normal patterns proved to have disastrous consequences not just in the Pacific region but around the world. In 1983,
for instance:

- Peru suffered-heavier-than-normal rainfall because of the
  warmer air which, in turn, caused flooding and landslides.

- Drier air over Australia resulted in the most disastrous brush
  fires of recent times, generally referred to as the Ash Wednes-
  day fires. In February 1983, 180 fires raged out of control in
  Victoria and South Australia propelled by 22 m/s (50 mph)
  winds, and heated by air temperatures of 40°C (104°F). More
  than 360 000 hectares (900,000 acres) burned, leaving 8,000
  people homeless and 72 killed. One fire-tornado is reported
  to have shot up 380 m (1,250 ft) in the air.

- Warm air currents shifted eastward and spawned six tropical
  cyclones, one of which devastated Tahiti.

- A northward shift in the jet stream resulted in increased rainfall in the Colorado River basin and in the Gulf States, causing
  floods, mud slides, and huge property loss.

- The eastward displacement of the warm Pacific air caused
  droughts in India and Sri Lanka and crop failures in Indonesia
  and the Philippines, with a heavy death toll.

- A displacement of global weather patterns caused droughts
  in Southern Africa that resulted in starvation, disease, and
  death. As a reminder that Mother Nature still has a few tricks
  up her sleeve, the government of a southern African country counted on a drought appearing with the 1997–8 El Niño.
  When it did not appear, the government found itself facing
  major economic problems. In anticipation of a drought, fewer
  crops were planted because banks tightened credit to farmers.
  When the rains unexpectedly arrived, it was too late to plant
  with the result that the crop yield was 20% below normal. The
  lesson from this is that the farther one is from the equatorial
  Pacific region, the greater the chances that other influences
  can modify an El Niño-based prediction.

Child Of Change

In the early sixteenth century, Spanish sailors traveling off the coast of Ecuador reported conditions of unseasonable rainfall and warm weather. Peruvian fishermen sailing from the port of Paita to Pacasmayo in the nineteenth century observed a warm coastal countercurrent that flowed south, devastating their fishing. They named it El Niño for the Christ Child, because it appeared after Christmas.

In the 1920s, Sir Gilbert Walker, the head of the Indian Meteorological Service, was studying monsoons, since he had earlier been asked to determine possible causes of the 1899 famine. He compared barometric (pressure) readings in Tahiti and Darwin, Australia, and noticed that when the pressure rises in the east, it falls in the west and visa-versa and called this the **Southern Oscillation**. He also realized that Asian monsoons were somehow linked to drought in Australia, Indonesia, India and parts of Africa, and to mild winters in Western Canada. Walker was ridiculed in some circles for suggesting such a global link in climatic conditions, just as Alfred Wegener had been disparaged in the same period for suggesting the continental drift theory that became the basis for plate tectonics, and led to an understanding of the origins of earthquakes. Both were later proven right. Jacob Bjerknes in the late 1960s had available observations of wind patterns above ground level and made the connection to the warm sea surface anomalies, the weak easterlies, and heavy rainfall in California to arrive at an understanding of El Niño.

The 1983 El Niño was clearly not an isolated event but a recurring one that can be traced back thousands of years. Archeological evidence points to a catastrophic flood in Peru in 1100 AD that was undoubtedly the result of an El Niño event. From tree-ring and ice-core studies, it appears that El Niño events occur every two-to-seven years, although anthropological studies on ancient garbage dumps suggests that there was a pause of 3,000 years, sometime between five and eight thousand years ago. Of all of nature's furies, El Niño has the most wide-ranging and disastrous meteorological consequences.

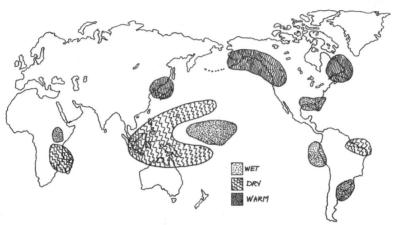

Fig. 13.3 El Niño in the Northern Hemisphere winter

Warm Brother, Cool Sister

ENSO, the *El* **Niño Southern Oscillation**, includes both the warm phase of the event and a cold phase known as **La Niña** (or El Viejo) that, in total, lasts from 3–7 years. Unlike the seasons, it is a highly irregular event and how and why it starts is still not perfectly clear. Since it involves the interaction of the ocean with the atmosphere, it is believed to be caused by a combination of either a disturbance to that system, such as an increase in thunderstorm activity in the Indian Ocean, or by the reflection of an oceanic wave from the western boundary of the Pacific. However, once it starts, the phases of the cycle are sufficiently well known to be able to predict its progression.

In a Northern Hemisphere winter (Fig. 13.3):

- The Peruvian-Ecuadorian-Chilean coast will be wetter than normal, benefiting farmers, but the oceans in those areas will be warmer than normal, hurting fishermen.

- The Southern United States and California will be wetter than usual as a result of the extra-moist subtropical air carried by the jet stream.

- The Northeastern United States, Western Canada, Eastern China and Japan will be warmer than usual.

- Australia, Indonesia, Northern Brazil and Southeastern Africa will be dryer than usual.

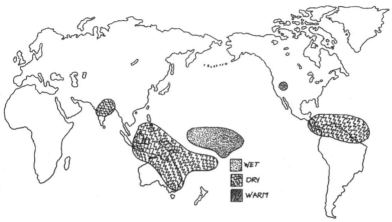

Fig. 13.4 El Niño in the Northern Hemisphere summer

In a Northern Hemisphere summer (Fig. 13.4):

- Australia, Indonesia, Central America and India will be dryer than normal.

- The Northwestern United States will be wetter than normal.

- There will be fewer Atlantic hurricanes due to the west to east winds that shear apart tropical disturbances before they develop into hurricanes.

Deep Lava

Searching for the trigger that sets in motion the chain of events causing El Niño disturbances, a geophysicist, Dr. Daniel A. Walker, has suggested that volcanism may be a factor. On the Pacific Ocean seabed there exists a rift, the East Pacific rise, which marks the boundary between the Pacific and South American tectonic plates (Fig. 13.5). It is a boundary along which the plates are separating[24] and through which red-hot magma flows as the plates separate. This magma immediately cools when coming into contact with the frigid seawater, spawning a plume of hot water that, according to Dr. Walker, may rise to the surface in the area near Easter Island. The consequent heating of the surface water could warm the air above and weaken the high-pressure cell that resides in that area of the Pacific Ocean. This high-pressure cell normally feeds the east-

---

24   See *Why the Earth Quakes*, Levy & Salvadori

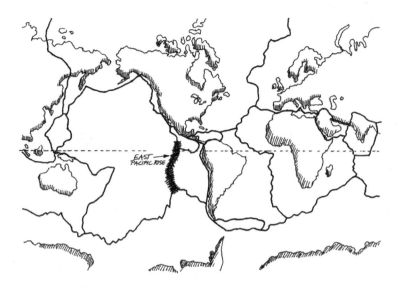

Fig. 13.5 East Pacific Rise along tectonic boundary

erlies that flow toward an area of low pressure near Australia. By weakening this airflow, the Southern Oscillation sequence of events is initiated.

However, volcanic activity along the rift is more or less continuous, so how could this play a role in initiating the Southern Oscillation? Reviewing the record of seismic activity originating from the rift, Dr. Walker noted that the periods of increased activity, and therefore increased flow of magma, correlated with the timing of El Niño events. Although this phenomenon may not be the only cause of El Niño, it may be a contributing factor. Many meteorologists believe that the main cause is a periodic oscillation of the coupled ocean-atmosphere system not unlike the sloshing in a bathtub, but on a giant scale in both time and space. Nevertheless, a volcanic origin could simply be another piece of the puzzle of the origin of El Niño.

14

Our Changing Climate
Global Warming and our Altered Future

> We are evaporating our coal mines into
> the air
>
> Svante Arrhenius, 1896

For millennia, growth in population and knowledge advanced slowly. Suddenly, in the early nineteenth century, population zoomed upward crossing the billion mark, and is now climbing above six billion (Fig 14.1).[25] By the end of the current century, it is expected that the world's population will grow to ten billion. Another development started in the late seventeenth century, with advancements in science and technology that set off an unprecedented industrial revolution led by the discovery of 97 new

---

25   In the middle of the eighteenth century, Benjamin Franklin observed that the colonies were doubling in population every 20 years, and predicted that in a century they would exceed England's population.

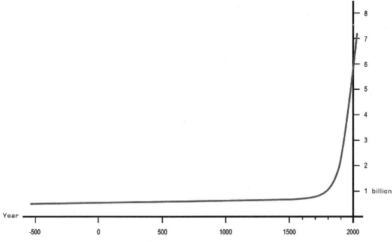

Fig. 14.1 World population growth

elements in a period of 250 years (Fig 14.2). The confluence of these two events resulted in a major expansion in the need for energy production.

Before 1800, wood was the principal source of energy in the world. In fact, 95% of the world's fuel was supplied by this source, with the remaining 5% being supplied by the muscle of man and beast. With the advent of the industrial age, the fossil fuels—coal, oil and natural gas—became the world's principal energy sources. As a byproduct of the combustion process, the production of energy

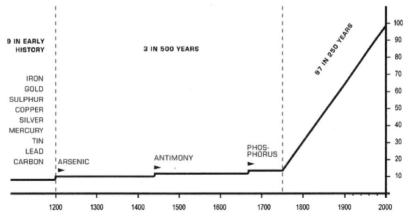

Fig. 14.2 Discovery of the elements

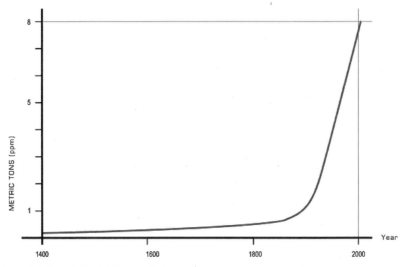

Fig. 14.3 Global CO_2 emissions

using fossil fuels releases carbon dioxide (CO_2) into the atmosphere. In the past, the carbon dioxide that is organically produced from human and animal waste could be absorbed by plants and the world's oceans, following nature's carbon cycle. Photosynthesis converts CO_2 and water that is soaked up by plants into carbohydrates and oxygen, by using solar radiation that is absorbed by the plants' chlorophyll. The world's oceans act as a sink for CO_2 through the process of solubility, consequently increasing the acidity of seawater.

Starting before the turn of the twentieth century, the enormous quantities of carbon dioxide discharged by our machines began to overwhelm nature's cycle and fled to the troposphere and stratosphere (Fig 14.3). There, these gases, (CO_2 as well as methane, nitrous oxide and others) enveloped our planet and, as in a greenhouse, permitted light to enter but prevented heat energy from exiting the troposphere. This resulted in what we know as **global warming**, causing a steady increase in the world's average temperature.

The Problem

For the past thousand years, the average temperature in the world has been relatively constant, though it had very slowly crept downward until the end of the nineteenth century (Fig 14.4). Since then, there has been a sudden, sharp, and continuing rise in temperature of 0.7°C (1.3°F). This is a seemingly small number when

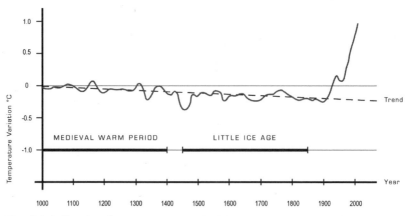

Fig. 14.4 Tracing the average global temperature

compared to seasonal temperature variations, but is startling when measured against the 5°C (9°F) rise that took place since the end of the last ice age about 10,000 years ago.

During this same millennium, the earth passed through the Medieval Warm Period that endured between the tenth and fourteenth centuries, when temperatures were about 1°C warmer than now. That was the time when vineyards flourished in England and citrus trees bloomed in north China, when alpine passes were open year-round, promoting trade between Italy and Germany, and when the retreat of the Arctic ice pack promoted the settlement of Iceland and Greenland. In contrast to this warm period, the earth passed through the Little Ice Age between the fifteenth and nineteenth centuries, when the average temperature was between one and two degrees cooler than today. At that time, the river Thames froze, permitting easy passage from one bank to the other, and Alpine glaciers descended, eroding valleys in their path. This was also the time when the sea ice expanded southward from the North Pole, isolating and ultimately dooming the Viking colony in Greenland. Meanwhile, in Europe, there were frequent crop failures with consequent famines as well as glacial landslides and avalanches, resulting in severe flooding in the spring snowmelt. By contrast, during this same period, some parts of the world experienced severe drought along with the cooling. For instance, California is believed to have suffered droughts lasting as long as 220 years. In the southwestern United States, the change from warm, moist conditions in the Medieval Warm Period to dry, cool weather in the Little Ice Age is thought to have caused the demise of the Anasazi Indian civilization.

Such warming and cooling periods have occurred naturally throughout the earth's history. Centuries-long cold spells such as happened during the Little Ice Age seem to recur every 1,400 to 1,600 years. These are then followed by centuries-long warming periods. This recurring cyclical variation in temperature has been cited by some as a reason for the increase in the world's temperature in the twentieth century. However, the suddenness of the present rise, coupled with increasing industrialization and a simultaneous increase in atmospheric carbon dioxide, suggests another cause.

The Human Imprint On Climate

Carbon dioxide is always present in the atmosphere. It is the food for all plant life and is, apart from water vapor, the principal greenhouse gas that occurs naturally and allows our planet to be habitable. Of course, the most common greenhouse gas is water vapor, which increases with increasing temperature. Like the glass roof of a greenhouse, these gases form an insulating blanket in the upper atmosphere, trapping heat by re-radiating heat energy back to Earth's surface. Without it, the planet would be a cold, inhospitable place with an average temperature of −32°C (−26°F) rather than the 16°C (60°F) that nurtures us. How delicate is the balance that permits life on Earth? If 25% less radiation were to reach Earth, its temperature would be lower than 0°C (32°F). With 25% more radiation, Earth's temperature would be higher than 30°C (86°F). Either of these conditions would be intolerable. Animals breathe in oxygen-rich air and breathe out carbon dioxide that is then absorbed by plants, which are irradiated by the sun and, through the process of photosynthesis, release oxygen back into the atmosphere. In the fall and winter, the process reverses as plants decay, releasing carbon dioxide. The world's oceans cooperate by absorbing about one third of human-generated carbon dioxide. It is a grand scheme of global recycling (Fig 14.5).

In the winter, the atmosphere holds more carbon dioxide than in summer, as if the earth itself breathes once a year. This cyclic event has taken place for hundreds of thousands of years while the average level of carbon dioxide remained relatively constant...at

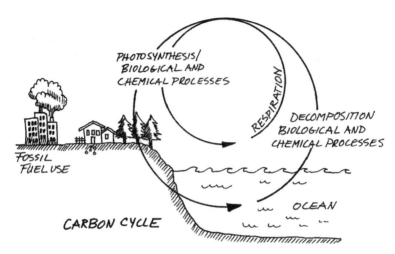

Fig. 14.5 The carbon cycle

least until the beginning of the twentieth century. Since then, the atmospheric carbon dioxide levels have risen steadily from about 290 ppm to over 380 ppm, a 31% increase, which is substantially greater than in any previous interglacial period and double the concentration in any ice age (as confirmed by analyzing and dating gases trapped deep in Antarctic ice). Dr. Charles D. Keeling was the first to measure carbon dioxide in the atmosphere. Starting in 1955 he would camp out at Big Sur in California, collecting air samples to measure the concentration of carbon dioxide. He had been recruited for this study by Roger Revelle, the director of the Scripps Institution of Oceanography, who had been one of the first scientists to suspect that carbon dioxide was responsible for the earth's warming. Over the years, Dr. Keeling plotted his results on what is now called the Keeling Curve, demonstrating the upward trend of carbon dioxide concentration (Fig 14.6).

Currently, the United States alone pumps 5.5 billion tons of carbon dioxide into the atmosphere each year. It would take a forest the size of Jupiter to absorb that much carbon dioxide. Even the world's oceans can't help, because their ability to absorb carbon dioxide diminishes exponentially as more is absorbed. Consequently, as much as 40% of it stays in the atmosphere and will reside there for centuries.

In the next fifty years the world's energy use is expected to double, led by the rapidly increased industrial development of China

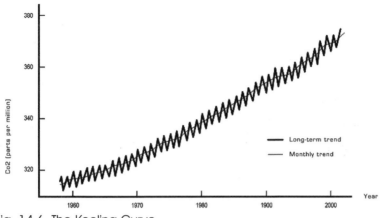

Fig. 14.6 The Keeling Curve

and India. China alone will burn enough coal, releasing sulfur dioxide and carbon dioxide as well as other noxious gases, to exceed
by 500% the emissions levels sought by the **Kyoto Protocol**.[26] This
results from the fact that a new coal-fired power plant is completed
every 7–10 days to fulfill the need for a growing industrial economy
but without the necessary equipment to remove pollutants from
smokestacks. The resulting polluting clouds that head east cause
acid rain, which is already affecting Japan and has even reached
the west coast of the United States. It is estimated that by 2025,
China alone will release twice as much carbon dioxide into the atmosphere as the United States. India, with a population that is expected to exceed China's by 2030, is right behind China in adding
to the problem.

To this day, the concentration of carbon dioxide in the atmosphere continues to increase, and within a relatively short time,
the continued use of fossil fuels will outstrip precious planetary resources and further intensify the global warming problem

To further exacerbate global warming, there are other greenhouse gases that are attributable to human activity including methane (from two billion farting and belching cows, sheep and goats, as
well as from rice paddies and sewage), a gas that is over twenty times
more effective than carbon dioxide in trapping heat; nitrous oxide
(from fertilizers), a gas that is 310 times more effective in trapping

26   The Kyoto Protocol, established in 1997, came into force in 2005 with ratification by 141 countries.
It establishes emission targets for each country with the objective of reducing emissions by 5.2% by 2012.
The United States has not ratified the agreement, saying that it would prove too costly.

heat; and industrial fluorocarbons such as CFC refrigerants and aerosols that, in addition to being highly effective in trapping heat, can linger in the atmosphere for as long as 50,000 years. However, carbon dioxide from the burning of fossil fuels remains the primary cause for unbalancing the global climate system.

How much longer can the planet deal with such enormous changes without rebelling?

The human imprint on current and future climate is incontrovertible and, as a panel of international climate experts agreed in 2001, "Most of the observed warming over the last 50 years is likely to have been due to the increase in greenhouse gas concentration." We have managed to make our presence felt on Earth—but to our future detriment. Lonnie Thompson, an Ohio State paleoclimatologist, has called this "a remarkable uncontrolled experiment." Even if we had stabilized emissions of greenhouse gases by the year 2000, the concentrations of carbon dioxide in the atmosphere are expected to have doubled from the pre-industrial level by the year 2100. By then, the resulting average temperature in the world will increase by about 3°C (5.4°F). (Without greenhouse gases, the earth's global temperature would actually have fallen during the twentieth and into the twenty-first centuries). The earth has not experienced this level of warming since the Climatic Optimum Period, from 7,000 to 3,000 years ago.

Consider the consequences of energy entering or escaping from what Buckminster Fuller, one of the most inventive minds of the twentieth century, called Spaceship Earth. In a cataclysmic event 65 million years ago, a 9-km-diameter meteorite struck the earth, raising a poisonous dust cloud into the atmosphere, causing the extinction of the dinosaurs[27]. Humans have now caused changes to the earth-atmosphere balance that will result in similar dramatic changes to life on Earth as we have known it. Consider that human activity in the past had impacts on local climate. For instance, the clearing of the Nile valley's lush forests resulted in today's arid landscape, and in the upper Yangtze basin, forest-cutting intensified the

---

27   In a possibly related event, a recent discovery extracted from 300-m-deep ice cores revealed that 55 million years ago, the Arctic was hot, with average temperatures of 23°C (74°F) and covered by palm trees and inhabited by alligator-like animals. This was followed by a cooling period of a million years that was initiated by a fern that sucked up massive amounts of carbon dioxide. This climate shift that originated with a heating cycle was caused by the release of greenhouse gases estimated to have been thirty times LESS than those currently being released into the atmosphere.

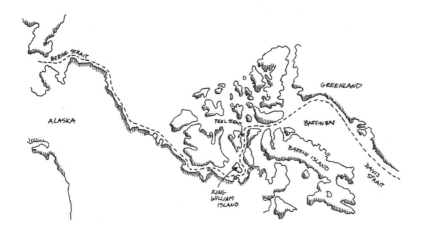

Fig. 14.7 The Northwest Passage

flow of the river, causing floods. None of these, however, was on such an unprecedented global scale as the situation we now face.

To visualize one consequence of global warming, consider the story of the Northwest Passage.

Northwest Passage

In 1506, the reigning pope divided the Atlantic and Americas between the two dominant maritime powers, Spain and Portugal[28]. This left commercial England seeking a Northwest Passage to the Far East that would bypass the territories under the control of her two enemies. Maps produced in the early sixteenth century depicted such a possibility. One was drawn by Girolamo de Verazzano, based on a voyage that his brother, Giovanni, took in 1524 as he explored the coast of North America. Since sailors had not yet penetrated the Arctic Region, the existence of such a passage was pure speculation, especially since the perennial Arctic ice extended down the coast of Greenland and penetrated all the way to the northern coast of North America. Nevertheless, the importance of a sea route along the coast of North America was spurred on by the need for trade between Europe, China, and India. By the late sixteenth century, maps identified Greenland and Arctic islands in the North Polar Sea—but no passage to the West. However, the inhabitants of the Arctic, the Inuits, had undoubtedly explored the northern reaches

---

28   Under the Treaty of Tordesillas, Pope Alexander imperiously divided the world. See also p.22

of the continent in their search for food and shelter. It is quite likely that they had traveled along the sea routes that formed the spine of the Northwest Passage.

Sir Martin Frobisher was the first English explorer to look for an approach along the northeast coast of the Americas. He was followed by others, including Henry Hudson. After discovering the bay that would be named after him, Hudson and loyal members of his crew were set adrift in a small boat by a mutinous crew and were never heard from again; the first of many who perished in search of the elusive sea passage. Enticed by prizes offered by the British government, Captain James Cook set out in 1778, northward from Hawaii, which he had just discovered, and passed through the Bering Strait seeking an access to the Northwest Passage from the Pacific Ocean. In the Arctic Ocean, he encountered only ice. He returned to Hawaii, where he found that he was no longer welcome, only a year after he had been treated as a god. The disenchanted natives stole one of his boats and, in retaliation, Cook took a Hawaiian chief as hostage. This enraged the natives, who surrounded Cook and his men. In the ensuing battle, Cook was stabbed to death, the same fate Magellan had met in the Philippines two hundred years earlier.

The next half century witnessed fewer attempts to discover the passage, primarily because England and France were occupied fighting each other, but also because of a change of attitude. It was no longer critical to discover the passage for commercial reasons, so interest shifted to exploration for scientific reasons.

Each expedition added more knowledge of the islands, straits, sounds, and inlets that make up the frozen regions north of Canada, and with each new geographic refinement, confidence increased that a passage would be found. On May 19, 1845, Sir John Franklin set out with a crew of 134 men in two ships, the *HMS Terror* and *HMS Erebus,* to sail the passage. Twenty years earlier, Franklin had explored the barren landscape of scrub and tundra between Kent Peninsula and Return Island, mapping almost half of the far-north American coast. He was confident that he had the information for a successful passage. Apart from his own reconnaissance, he relied on a map prepared by James Ross in 1831, showing a land connection between King William Island and Boothia Peninsula. There is actually no land bridge between the two, and an isthmus separates them, an unknown fact that was to prove disastrous to the expedition. Starting from Lancaster Sound, Franklin intended to sail through Barrow Strait, then down along the coast to the Bering Strait (Fig 14.7). Passing the west coast of Greenland, the expedition exchanged greetings with whalers on their way home. This was the last contact his countrymen had with Franklin and his men. From evidence gathered from some of the forty search parties that were sent to find out what happened, the following sequence of events was reconstructed.

The first winter, Franklin established an encampment on Beechey Island, off the southwest coast of Devon Island. During the winter, three of his men died of unknown causes (although one may have had pneumonia). When the ice broke in the late spring, the expedition sailed up Wellington Channel, across Barrow Strait and down into Peel Sound. As the sea ice thickened in winter, the ships became wedged in Victoria Strait, where the multiyear ice pack enters from the Arctic Ocean. By following the west side of King William Island, Franklin was acting on the information obtained from James Ross's map. Had he taken the east coast, he would have encountered only annual ice that would have broken up in the spring. Instead, his ships were held as in a nutcracker, unable to move as the ice screamed like a banshee against the hulls. Such was the force of the ice that it could lift a ship out of the water or roll it on its side. By the second winter, Franklin and twenty-three of his men had died. The cause is still a mystery, but it is suspected that food from improperly sealed tins caused botulism and that lead from the tins may also have contributed to a slow poisoning of the men. In the spring of 1848, the survivors, under the command of Captain Francis Crozier, abandoned the ships and set out across the sea ice to the coast of King William Island, where they left a message in a cairn. It indicated that the captain intended to lead the 104 remaining men 420 km (250 mi) to Back's Fish River. During this trek, many men apparently died, and Inuits who found their bodies recovered relics that were eventually turned over to later explorers. On the south coast of King William Island, Crozier, with only forty desperately starving men, encountered four Inuit families from whom they bought some seal meat. Some time later, the last thirty of Crozier's men were found dead near the mouth of the Fish River. A boat was abandoned on the beach, indicating that they may have hoped to head upriver to Fort Providence 830 km (500 mi) away.

The expedition's journals were found, waterlogged, wind-scattered; an indecipherable mess. There was evidence that the men may have been trying to kill and eat the first snow geese arriving from the south. Instead, the campsite, with some butchered and partially eaten corpses, illustrated how Franklin's glorious adventure ended in cannibalism. A disbelieving widow, Lady Franklin, could not accept the fact that British naval men could have been reduced to behave in such an animalistic manner. Her anger was directed at the messenger of the news, the great arctic explorer John Rae. Rae had led the search party that, in 1854, had discovered the shocking truth after interrogating the Inuits at the base of the Boothia Peninsula. Lady Franklin enlisted the support of Charles Dickens to write a series of articles that had the effect of destroying Rae's credibility. In the end, both her husband's and Rae's reputations suffered immeasurably from her vengeful act.

Roald Amundsen, a Norwegian explorer, finally succeeded in traversing the full extent of the Northwest Passage in a three-year expedition in 1903.

Most of the early exploration of the passage occurred during the Little Ice Age (Neo-Boreal) that extended four hundred years from 1450. During the warming period after that, by the latter half of the twentieth century, the ice that had clogged the arctic region had been dramatically reduced by over 40%. This trend accelerated so that ice that was measured in the region of 72°N to be 3 m (10 ft) thick in the 1970s had become wafer-thin twenty years later. Since bright ice reflects more than half of the sunlight that reaches the earth's surface and dark water absorbs 90% of the sun's energy, less ice leads to increases in temperature that, in turn, lead to more melting. In the atmosphere above the Arctic, another phenomenon contributes to the destruction of the ice pack. When air pressure increases over the Arctic, it decreases over a doughnut-shaped ring at 40°N and reverses at a later time. This **Arctic oscillation** used to shift randomly but recently favors low Arctic air pressure and consequent stronger westerly winds that tend to break up the ice pack.

As a result of the warming in the Arctic region, it is expected that in the early decades of this century, ships will be able to cross the Northwest Passage unassisted for up to three months a year, and more if assisted by an icebreaker.

Melting Ice

Glaciers cover 5% of the state of Alaska, serving as the source of most of its rivers and containing three quarters of its fresh water. Since about 1900 these glaciers have been retreating, and have recently been observed to be shrinking at double the rate of the last decade of the twentieth century. The retreat of glaciers is being noticed all over the planet, from the Alps to the Andes to the Himalayas and even Kilimanjaro in Africa. As a consequence, the areas that rely on water-melt from the glaciers for fresh water will eventually wither and dry up.

Not only had the Arctic ice decreased in the latter half of the twentieth century, but the Greenland ice sheet also started melting. In five recent years, South Greenland lost 5 cu km (2 cu mi) of ice. Usually, a warmer atmosphere holds more moisture, which leads to heavier winter snowfalls that are then compacted into more ice. Yet warmer temperatures tend to melt the ice at its edges and at lower altitudes. Since the Greenland ice sheet holds 6% of the world's fresh water, melting it can radically change the salinity of the ocean, with serious consequences to the currents circulating in the oceans. Typically, warm, fresh water moving up the Atlantic Ocean from the tropics encounters saltier, heavier water south of Greenland that then sinks deep below the surface and travels south to the coast of Antarctica, where it splits into two branches: One travels north into the Indian Ocean and the other skirts the eastern shore of Australia on its way to the North Pacific. Both branches then rise and travel back to the Atlantic Ocean (Fig 14.8). Appearing like a continuous conveyor belt, this **thermohaline circulation**, so called because it is driven by differences in the ocean's temperature and salinity, moves at a snail's pace of 0.1 m/sec, with the consequence that the cooled water sinking south of Greenland can take a thousand years before seeing the sky once more.

Fresh water flowing from the melting ice could weaken and slow this circulation with global climatic consequences because of the ocean-atmosphere link (as discussed in Chapter 13: El Niño). This conveyor belt has shut down many times in the past and done so relatively quickly, with resulting altered climates that 8200 years ago lasted a century, and during the Younger Dryas time (12,700 years ago), lasted a thousand years. Such an event can lead to the

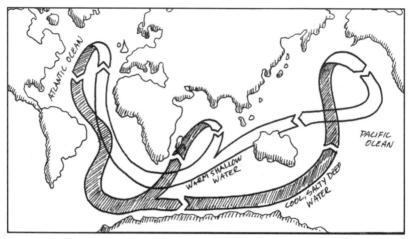

Fig. 14.8 Thermohaline circulation

cooling of England, Northeastern North America, and Europe by up to 4°C (6°F) since these regions would be denied the warming of the tropical current. A significant slowing of the current that caries warm water to Northern Europe has already been observed by the UK National Oceanography Centre. Compared to readings taken in 1957, a 30% drop in flow was measured in 2006. Other possible consequences of the slowing or stopping of the thermohaline circulation include increased temperatures in parts of Australia, South America and South Africa; intensified winds and winter storms in Western Europe and the North Pacific; and drought in agricultural and water resource regions of Europe and Eastern North America.

Solar Cycles

The sun, which warms and nurtures us, does not do so at a perfectly constant rate. Periodically, its surface is ravaged by storms that are seen as sunspots—darker splotches on the surface that indicate magnetized regions with cool gases. As some of these regions with opposite polarity meet, magnetic energy is released as heat, and giant flares spewing up tons of solar material shoot out from the surface. At such times, the sun appears to burn brighter, sending more of its warming rays to the earth. Lasting between 8 and 15 years, with an average of 11 years, sunspots are linked to changes in the earth's weather. First, there is an increase in warmth that may account for almost half of the warming that has occurred since 1850. (*Greenhouse gases are needed to account for the balance of the in-*

creased warmth.) Second, the sun's increased brightness leads to an increase in ultraviolet radiation, which produces more ozone in the upper atmosphere and is linked to shifts in normal storm patterns. It has been observed that winter storms that sweep across the Mediterranean during periods of solar maximums (the last in the year 2000) move northward about 600 km (400 mi). Finally, solar winds of ionized hydrogen and helium particles block cosmic rays that are known to enhance rain formation in clouds. These are the same solar winds that stream past the earth at 1.5–3 million km/hr (1–2 million mph) and cause the aurora borealis, or northern lights.

A drop in sunspot activity that led to cooling the earth by 1°C (2°F) in the period from 1640 to 1720 is an example of the solar effect. During that period, glaciers expanded and winters in Europe were longer and more severe (the Little Ice Age).[29] The influence of the sun cycles on the earth's climate is undeniable and has been used by naysayers of the human imprint on climate[30] as justification for inaction in the efforts to decarbonate the global energy supply. Some have argued that it is perfectly all right to allow the earth to become warmer, since in past warm periods, mankind prospered more than during cold periods. Yet the evidence of the potential damage resulting from global warming involves not just an increase in temperature but a catastrophic disruption of the global climate system that is predicted to lead to decades-long fundamental changes.

Other factors cited by those who deny that humans are causing global warming are based on the premise that the earth has been subject to temperature variations over the millennia due to natural cyclical variations such as those identified by Milankovich.[31] But this runs counter to the major findings about global warming that have been generally agreed upon by scientists:

- The CO_2 in the atmosphere during the last ice age was 180 ppm and, after the glaciers retreated, climbed to 280 ppm. The concentration remained relatively constant after that time but has now climbed to 380 ppm and is projected to go

---

29   Dr. Gerald C. Bond (1940–2005) found that glaciers have melted every 1400–1500 years. He theorized that variations in sunspot activity may be responsible.

30   Anthropogenic climate change

31   Milutin Milankovitch, a Serbian mathematician, established a cyclical relation between ice ages and planetary variations: Every 100,000 years the earth's orbit stretches from near circular to elliptical; every 41,000 years there is a variation in the tilt of the earth of about 3 degrees; every 19–20,000 years there is a wobble in the earth's rotation.

to 500 ppm before the end of this century. These concentrations are greater than those experienced in any interglacial period. Since CO_2 remains in the atmosphere for up to 200 years, any attempt at mitigation can only slow, not stop, the consequences of global warming during this century.

⇒ As for temperature, there has been a 0.7°C (1.3°F) increase since the end of the nineteenth century above what had been a relatively constant average temperature for the past thousand years. In fact, this sudden rise countered what had actually been a slowly dropping average temperature during the last millennium. By the end of this century, the average temperature rise is expected to reach 3°C (5.4°F).

⇒ With rising temperatures, the habitat of plant and animal species has migrated northward by about 50 km per decade in the Northern Hemisphere, forever changing the ecological balance in these regions. In England, for instance, spring now arrives a week earlier than in the 1970s.

⇒ The Antarctic is losing 150 cu km of ice per year, uncompensated by yearly snowfall. If all the Antarctic ice melted, sea levels would rise 65 m (215 ft). Since oceans absorb as much as 90% of solar energy and ice reflects 80%, more open water implies higher temperatures.

⇒ Glaciers throughout the world are retreating.

⇒ The Greenland glaciers are melting, with a doubling of the flow of ice in the past decade. This has the potential of ultimately raising sea levels by 7 m (23 ft).

⇒ Over the last century, seas have risen 150–200 mm due to a combination of expansion (resulting from increased temperature) and fresh water added from melting ice. It is currently rising at the rate of 2.8 mm per year.

⇒ The equilibrium existing between gases absorbed and released by the oceans has been disrupted with the result that the increase in dissolved carbon dioxide that produces carbonic acid, has caused a 30% increase in the oceans' surface acidity,[32] damaging coral and sea shells and unbalancing the

---

32   Acidity is measured by pH where 7 is neutral and lower numbers refer to increased acidity. There has been a decline in the oceans' surface pH of 0.1 from a more alkaline range of 7.8–8.5. Since this scale

food chain. This is in addition to the problems caused by the increase in ocean temperature that also damages marine life. Although the oceans have a buffering capability through the dissolution of marine shells that lie on the bottom, it may take thousands of years to do so because the slowness of the deep ocean thermohaline circulation (see p. 151).

⇒ Permafrost is melting in Siberia and Alaska, with the consequence that methane and carbon dioxide that had been trapped within it since the end of the last ice age (over 10,000 years ago) will be released over the next few decades. Potentially, up to 800 gigatons of carbon could be released from the melting permafrost, compared to the annual human output of about 7 gigatons. This occurs through a self-perpetuating mechanism by which increased heating causes release of greenhouse gases, which in turn cause increased heating. Already, structures that were built on permafrost have begun to sink, including the piers supporting the pipeline bringing oil from Alaska to the lower forty-eight states. As the permafrost melts, it also creates thaw lakes with decomposing organic material releasing bubbles of methane.

⇒ The ten hottest years in historical record have occurred since 1990.

⇒ The most costly natural disasters have occurred since 1988. Average wind speeds in storms have increased 50% in the last fifty years and ocean temperatures are increasing, leading to more intense storms. Among them was Hurricane Katrina, the costliest and one of the most deadly storms ever to strike the United States.

Hurricane Katrina

Hurricanes, typhoons, and cyclones originate over warm oceans north and south of the equator. On August 24, 2005, one such storm started out as a tropical depression east of the Bahamas. As it began to swirl and draw increased strength from the warm waters, it turned into a tropical storm that was named Katrina. With winds of 130 km/hr (81 mph), making it a Category 1 hurricane, it struck the southeast coast of Florida north of Miami. As it passed over the land it weakened, but once it moved across the warm waters

is logarithmic, such a change represents a 30% decline.

of the Gulf of Mexico, the storm once more gained strength, becoming a Category 2 hurricane. It also turned to a northwesterly
direction, aimed directly at the Mississippi and Louisiana coast. By
August 28, it had ballooned into a Category 5 storm with winds of
280 km/hr (175 mph). On the following day, when it hit the coast,
it had weakened to a still-deadly Category 4 storm with winds of
230 km/hr (144 mph). More significantly, it was an unusually huge
storm, with hurricane winds reaching out 190 km (120 mi) from its
center, and was thus able to devastate extensive coastal areas all the
way from Louisiana to Alabama.

The devastation included the catastrophic destruction of the
friendly and festive city of New Orleans, with its unique blend of
the old and new world. This city, 80% of which lies below sea level,
is particularly vulnerable to a storm surge as it lies between the
Mississippi River and Lake Pontchartrain (Fig 14.9). Levees along
both of these boundaries had been built 7 m (23 ft) high following
the disastrous 1927 Mississippi River flood (see p. 124). Hurricane
Katrina's right-front quadrant, containing the strongest winds, was
predicted to produce a storm surge that could overtop the levees.
Although this did not happen, a more dangerous problem developed, as several weakened sections of levee were breached, allowing
water to enter and flood the city. The combination of wind and water destroyed large parts of the city and caused the deaths of almost
two thousand people, with many still missing after the levees had
been repaired and water had been drained from the flooded areas.
More than a year after the event, it is not at all clear that the city
will ever regain its unique character and vitality.

There have been hurricanes throughout history in this region,
but the intensity of this storm was unique. Since a storm's intensity is
directly related to the warmth of the ocean that provides its strength,
an increase in ocean temperature links it to one consequence of
global warming. Worldwide, there do not appear to be more or fewer
storms on average, but the intensity of individual storms has definitely increased in the last half century as evidenced by Hurricane
Katrina, as well as typhoons that struck the coast of China in 2006.

***Knowing the facts of global warming, we must now ask, what
are its consequences?***

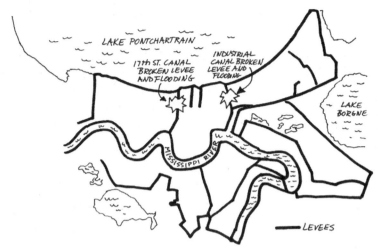

Fig. 14.9 New Orleans levees

The Future Climate

What can we expect during the next hundred years? As Niels Bohr said, "Making predictions is a very dangerous thing, especially if they are about the future." Nevertheless, there are numerous uncontested consequences of global warming. Because of a lag in the response of global systems to actions taken today, even if there is an immediate reduction in greenhouse gases released into the atmosphere, many of these predictions will undoubtedly come to pass.

- Global temperature will increase by about 3°C (5.4°F). This is estimated from the fact that since the last ice age, warming was equivalent to 7 watts/sq. m with a temperature increase of 5°C, whereas the predicted greenhouse warming is about 4.5 watts/sq. m. Warming, though, will not be uniform. It is anticipated that nights will warm more than days and result in an increased cloud cover. There will be more warming in winter than in summer. Also, hot nights will be more prevalent than cold days.

- As a result of the temperature increase, sea levels will rise by about 500 mm (20 in), partially due to melting ice and partly from the expansion of the oceans as a result of the increase in temperature. (Even today, the water level in New York Harbor routinely rises by 180 mm (7 in) as winter turns to summer). Islands in the Southwest Pacific Ocean, the Indian Ocean, and the Caribbean that are close to sea level will be engulfed and drown. The Polynesian islands of Tuvalu have already been partly evacuated because of the rising waters, and are expected to be completely underwater by 2050 as a result of a predicted 320 mm (12.5 in) rise in the ocean level. In the rest of the world, the submergence of seacoast territory could displace as many as seventy million Chinese, thirty million Bangladeshis, and thousands of Americans as the sea swallows up parts of Florida and Louisiana.

A worst-case scenario would involve the melting of the west Antarctic ice sheet, which would result in even more disastrous consequences, with a sea level rise of as much as 6 m (20 ft). As a warning of such a possibility, in 2002, a massive ice shelf, the size of the state of Rhode Island, collapsed into the Antarctic sea over a period of a month, opening the way for increased ice flow off the continent. Of course, before this happens, the Greenland glacier would long ago have melted! The danger lies in a feedback phenomenon that causes a rapid and irreversible rise in sea level as the temperature that existed when the sea was 4–6 m (13–20 ft) higher is approached.

≡ Storms and floods will become more severe and deadly. In general, weather will tend to be more extreme, with heavier rain and snowstorms and longer dry spells. There has already been a 20% increase in the proportion of precipitation that falls as downpours in the last century.

≡ Hurricanes will become more destructive. The frequency of tropical cyclones may be greater in the western Pacific and fewer in the Atlantic, and may thus balance out globally. However, as described in the discussion of Hurricane Katrina, their intensity may increase due to warmer seas that feed their energy.

≡ Deserts will expand and periods of drought will lengthen. California is particularly vulnerable to the possibility of decreased precipitation. Increases in large forest fires caused by hotter and dryer seasons have already occurred more frequently in both the western United States and Australia.

≡ Glaciers, nature's natural reservoirs, will continue to recede. It is estimated that the glacier known as the Snows of Kilimanjaro will disappear by the year 2015 and some South American glaciers and Alpine glaciers will completely vanish within the next quarter century. Even in the mighty Himalayas, snow-and-ice cover has shrunk by over 30% since the 1970s. Melting glaciers in the high mountains of the world will threaten the water supplies for drinking water, irrigation (threatening food supplies) and hydropower. This will have devastating effects in poor countries such as Peru and Bolivia, which have no way of replacing the lost food supplies, water, and power.

- In southern Canada and the northern United States, the ratio of snow to rain will decrease while there will be an increase in the overall quantity of precipitation.

- There will be a decrease in precipitation in the tropics and subtropics, contributing to the expansion of the deserts in those regions.

Only time will tell which of these predictions will come to pass, but there is no question that for the first time in human history, man has caused a significant disruption of the natural climate system, resulting in potentially devastating environmental consequences. *There will be consequences!*

The Solutions

Having caused the problem, humans now face the task of finding a solution, if indeed there is one. We will certainly have to adapt to the consequences of a warmer earth. The process that began over a hundred years ago cannot be stopped, and even if we introduce changes to reduce carbon dioxide emissions today, these may only slow the process, delaying some of the ultimate consequences. Nevertheless, we can and must take steps to protect future generations, perhaps following Buckminster Fuller's assertion, "Do not fight forces, use them!" Considering the fact that vehicle exhausts from cars, trucks, planes and ships contribute about 14% of emissions, deforestation contributes 18%, and power generation contributes 25%, it is logical to tackle the worst offender first. It definitely makes sense to reduce emissions from vehicles and to stop deforestation, but reducing emissions from power generation would contribute the most to attenuating global warming. Here are a few ideas:

Biomass. Renewable agricultural products already convert CO_2 through photosynthesis and then decay by aerobic (releasing heat, water and CO_2) or anaerobic bacterial processes, releasing methane. As wood was used in the past, ways of using this potential energy source with the aim of replacing 10% of fossil fuels can be brought about by:

- Developing plantations with rapidly growing trees such as eucalyptus and sycamores.

- Recycling to collect farm waste, urban waste, wood manufacturing waste, and sewage—capturing the methane to use as fuel.

- Ocean farming to collect sea kelp.

One application is in the production of ethanol, which is today one of the principal fuel sources in Brazil, making use of their lush environment. It is an ecologically preferable fuel, but does not solve the long-term CO_2 emission problem. There is an approximate balance between the CO_2 absorbed by the growing plant and that released by burning the produced fuel. However, additional CO_2 is released in producing the fuel.

Nuclear fission was once thought to be the answer to our long-term energy needs, at least before Three Mile Island and Chernobyl awoke us from our innocent belief in a problem-free nuclear energy future. Those areas of the world that do not have native fossil fuels have been very aggressive in promoting nuclear power generation. While nuclear fission provides about 20% of the US capacity, Europe relies on it for more than half of its power needs. Yet the major unresolved problem with nuclear energy is the disposal of the spent fuel, which still remains highly radioactive although it no longer functions to fuel plants. After more than forty years of controversy, it is not clear whether disposal should take place by encapsulating, burying, or storing. It has become a classic "NIMBY" (not in my back yard) situation where no one wants the disposal to take place near his or her home. However, there remains the dream of a nuclear future that necessitates the development of breeder reactors and eventually fusion reactors—still a possibility before the end of this century.

Hydroelectric power is naturally limited by the availability of suitable dam sites and by the potential environmental problems they create (see Appendix B). Nevertheless, China, for example, obtains 17% of its power from hydro, a figure that is projected to grow to 40% now that the massive Three Gorges Dam is complete, and with the completion of the power station by 2015.

Solar. The largest essentially untapped source of non-polluting energy is that associated with the sun. The potential for solar energy is immense. For the United States alone, the total available is 500 times the current consumption rate. It is also 1,500 times greater than the available geothermal capacity and 60,000 times greater than the total available tidal energy. In order to take full advantage of this potential, solar collection systems would need to partially cover the earth, a clearly impractical solution. The source, however, is virtually permanent with no degradation of availability. Great strides have been made in the development of collection devices from the direct photovoltaic silicon-based panels to a heliostat field of concentrating mirrors aimed at a collection tower that feeds a thermal storage device. Placing collection devices in the desert areas of the world would meet a large percentage of the world's energy needs.

Wind. Windmills have always inspired the romantics; Don Quixote battled "thirty, lawless giants with many arms...." (See Appendix A). Holland has its four-arm windmills and Greece its roller-reefed sail windmills. They have been used over the centuries for grinding grains and lifting water, but electric power generation is a new role for windmills. Block Island, off the state of Rhode Island, today receives most of its power from windmills, and all over the world wind farms are providing more and more power. William Heronemus of the University of Massachusetts has proposed a network of off-shore wind power stations with 60-m (200-ft) diameter wind turbines. A forest of 300,000 windmills has been proposed for the great plains in the central United States. It has been estimated that 75% of the two trillion KW presently consumed in the US could be generated by wind. Although this source is clean, the construction of new windmills is also limited by NIMBY campaigns and many proposals are stalled for this reason.

Solar Chimneys. The engineer Jörg Schlaich has explored the possibility of developing solar chimneys for power generation. This unique means of solar energy generation makes use of the natural draft that exists between a ground-based, covered collection area surrounding a chimney and the top of the chimney. A turbine installed at the base of the chimney serves as the generator. Such installations are very practical in hot climates and could be built in deserts with minimal environmental impact. An experimental installation in Spain confirmed its potential and served as proof of concept. This century may very likely see the proliferation of wind turbines and the introduction of solar turbine generation becoming a reality.

As we await the arrival of many of these energy sources, we can also continue to develop technology to further lessen our dependence on fossil fuels by developing hydrogen-based internal-combustion engines and fuel cells that have mainly clear water as a combustion product. The hydrogen could be obtained by electrolysis using power obtained from wind or solar sources or from coal with carbon sequestration using ceramic membranes for hydrogen separation. Ultimately a hydrogen-based energy system, with water as the waste product instead of carbon dioxide, may become practical. It is not a solution now, because creating the hydrogen fuel currently requires more energy than that in the resulting fuel. Furthermore, the energy used in creating the hydrogen fuel is currently a fossil fuel that spews carbon dioxide into the atmosphere and will not be non-polluting until wind or solar energy are its source.

Until ways are found to overcome those drawbacks, air and water offer the best alternative (other than nuclear) sources of energy replacing polluting carbon. In addition, there is a need to fundamentally change attitudes away from consumption and toward **conservation and sustainability** while at the same time reducing carbon dioxide emissions. In a coal-based economy such as China's, this means converting to coal gasification and sequestering the offensive gases by pumping them into the earth.

Such changes will undoubtedly prove to be difficult, but we must face up to the fact that even the most daring solutions to this problem of global warming will not bring immediate rewards. As has been stated earlier, changes have been set into motion that cannot be reversed quickly. It is like trying to stop a heavy freight train, knowing full well that even once we apply the brakes, the train will still travel a long distance before stopping. In a society accustomed to instant gratification, the repair to the problem we have created will need a decades-long effort. Yet this is the challenge that we must accept: to take whatever action is necessary in the short term to reduce greenhouse gas emissions in order to protect our children and their children from the damage that we, and our parents and grandparents, have caused. Switzerland, that small country in the center of Europe, has established a goal of reducing its per-capita energy consumption to 2,000 koe (kilogram of oil equivalent), a reduction by almost half of its current consumption. Were that to be established as a worldwide goal, it would go a long way toward alleviating the damaging consequences of global warming and provide time for the development of long-term solutions[33]. The reality remains that cutting emissions by half will only double the time for the damaging effects to take place; it will not reverse the trend.

What is needed is a global commitment to a sea change that has as its goal a zero tolerance toward damaging emissions. The success of the worldwide effort in the 1970s to reduce and reverse the emissions of chlorofluorocarbon (CFC) that were causing a hole in the ozone layer serves as an example of what can be accomplished with concerted worldwide political action.[34] Dr. James Hansen, chief of the NASA Institute for Space Studies, a leading proponent of government action, has suggested for consideration, a carbon tax as well as international cooperation that was an objective of the Kyoto Protocol. He points to the danger of inaction by noting that a 1°C increase in worldwide temperature is manageable but a 3°C increase results in irreversible consequences. When will we reach this tipping point?

We are almost there!

What can governments do to stabilize the global climate when improving efficiency and expanding renewable energy is clearly not enough? It is a global wartime challenge. At the start of the Second World War, the United States was able to convert a peacetime in-

33   In 2006, the state of California mandated a cut of 25% in heat-trapping emission by the year 2020. Also, Sweden is commited to becoming the first oil-free economy by 2020 and even China and the European Union have goals of obtaining respectively 16% and 12% of their energy needs from renewable resources.

34   The 1987 Montreal Protocol banned the production of CFCs and resulted in a reduction in the ozone hole over the poles.

dustrial economy to full war production in the space of less than
a year. It also funded, organized and accomplished the development of an atomic weapon in less than five years. A similar techni-
cal challenge faces us today except that it is on a global scale. All
nations must commit to engaging to solve the problem—Kyoto is
only a feeble beginning. A dramatic and costly timetable must be
established to achieve what may seem like impossible reductions in
carbon emissions to pre-industrial levels. To put this in a context,
the average per capita emission of carbon has changed little over
the years while the population since 1900 has increased by 700%.
To achieve our goal we must therefore reduce worldwide emissions
by at least 700% of the 1900 levels. This implies that the largest pol-
luters, such as the United States, which have five times the average
per-capita emission rate, face a major challenge. There is no time
for polite discussion when dramatic, politically painful and costly
action is needed.

Until the governments of the world can be convinced to take
drastic measures to curb carbon emissions, we, as responsible citi-
zens, can only take relatively small steps and make efficiency, sustainability and conservation our objective. As Aldo Leopold, a con-
servationist, wrote in 1946, "That the situation is hopeless should
not prevent us from doing our best."

We can all contribute.

It is sad that most of us will not live to see the benefits of our
efforts but for once in the history of human existence, we must act
to benefit, rather than harm, future generations.

A

Appendix A
The Power Of Wind

> They set the mast in its socket in the cross
>   plank,
> Raised it, and made it fast with the
>   forestays;
> Then they hoisted their white sails aloft
>   with ropes of twisted oxhides.
> As the sail bellied out with the wind, the
>   ship flew through the deep blue water,
> And the foam hissed against her bows as
>   she sped onward.
>
> *The Odyssey*, Homer

The first people to appreciate and use the power of the wind were the ancient mariners. As early boats used the muscle power of rowers to propel them, using the wind must have seemed like a miraculous labor-saving device. In the Mediterranean basin, the Egyptians invented the sailboat around 3200 BC to provide the motive power needed for long voyages and fishing in far offshore waters. These first sailboats, with wooden hulls and triangular sails, are not so different from the felucca that still travel along the Nile. The Minoans and later the Phoenicians improved on the early designs, and eventually built large fleets that traveled around the Mediterranean. As long as the wind blew from behind, these early sailboats worked well, but lacking a rudder, they needed to be steered with oars from the stern. The Chinese were the first to overcome this shortcoming by using a rudder fixed to the stern, and introducing a keel or centerboard to stabilize the craft laterally. These design features were introduced into the junks that still roam the South China Sea. Unlike the triangular lateen sails that evolved along the Mediterranean coast, the junk has square sails stiffened with horizontal bamboo battens. Over the centuries, wind-powered craft crisscrossed the world's oceans in voyages of discovery, as warships expanding and protecting empires and as commercial vessels moving goods from one end of the world to the other. The development of the steam engine in the nineteenth century slowly rendered them superfluous.

Searching for another labor-saving device, diverse civilizations came upon the idea of the windmill. Whether used to pump water for irrigation or to mill grains into flour, windmills appeared over 2,500 years ago in China, Persia, and the Mediterranean basin. Rather than the horizontal axis machines we are familiar with, these early windmills used vertical axis rotors. One of the earliest known examples of such a machine was the Persian windmill built in Seistan, near the Afghanistan border. It consisted of a vertical paddle wheel surrounded by a curved wall on one side and a straight wall on the other. Wind blew through the opening between these walls, driving the paddles (somewhat like a revolving door). In the second century BC, Hero of Alexandria built a horizontal axis windmill to power an organ. But the major development of windmills did not take place until the eighth century, first in the Mediterranean region and then in Europe after being introduced by the returning crusaders. In the thirteenth century, postmills appeared. These were boxy structures, balanced on posts, with the horizontal axis rotors mounted on them. To face into the wind, the whole mill had to be turned manually, using a long pole attached to the top of the housing. Within a hundred years, smockmills were developed—structures on which the top housing rotated on a fixed base. In 1745, Edmund Lee invented the fantail, which turned the rotor assembly automatically into the wind. By that time there were over twenty thousand windmills in England and the Netherlands. These were used for milling grain, milling paint pigments, oil, and glue (from cowhides and animal bones), hulling and fulling (to shrink and thicken cloth), sawing timber, and processing paper. The sails on these early mills were of canvas and needed to be furled in heavy winds to prevent their being torn apart. Don Quixote aimed his lance at such windmills as he engaged in battle against "thirty lawless giants with many arms." Sail-rotor windmills still grace the scenic landscape of the Greek islands, and the Dutch countryside is still dotted with the windmills that we see represented in Old Master paintings.

Clifton Boyd, a miller in Rhode Island, ran his family's eight-sailed windmill in the early twentieth century. "I was running that mill when I was 14 years old," he wrote. "It was a good deal like sailing a ship. You had to keep an eye on the weather, especially in winter, and we ran it winter and summer. When the wind changed, you stopped the mill right away, took in the sails, and turned the top to face the wind again. You couldn't run it in a thunderstorm because the wind would change too quickly. You had to let the brake on easy. Sometimes if you got caught in a very high wind you could put the brake on and the mill would keep right on going until the gust stopped. Then you would reef in the sails a little more. You could reef our sails standing on the ground, but with eight vanes, eight sets of sails, it was a long job."

In the nineteenth century, wind-driven pumps were used extensively in the developing western part of the United States, and over two hundred thousand are still used today.

After the First World War, many of these were converted to generate electricity to charge batteries used to power the first crystal and tube radios—but with only moderate success. When Marcellus Jacobs developed an efficient three-blade propeller generator, he sold hundreds of thousands, many of which are still used today for both pumping water and generating electricity in remote regions of the world.

Such wind-driven generators were even used on ships. In 1883, the *Fram*, a 128-foot, three-masted schooner, explored the Arctic sea. It was an extraordinary ship, with a four-foot thick oak hull reinforced with iron to resist the pressure of the ice. More unusual was the fact that it had a windmill to power electric lights, which had only recently been invented. But the use of wind power to generate electricity was treated as an expensive novelty until the late twentieth century. A determined engineer who saw its potential led a resolute effort to demonstrate its commercial feasibility.

Grandpa's Knob

During the First World War, Palmer Cosslett Putnam (1900–1984) served in the Royal Air Force before going back to college. He then graduated from MIT as a geologist with an interest in volcanoes. After a time exploring volcanoes in Central America and serving as a geologist in the Congo, he joined the family publishing house, GB Putnam, in 1930, and two years later, became its president. He was clearly not destined to become a businessman, as he had to declare bankruptcy in 1934. His interests lay elsewhere. After building a house on Cape Cod, he became greatly dissatisfied with the cost of utilities. Standing on the shore, he marveled at the constancy and strength of the wind, and considered building a wind turbine to provide power for his and his neighbors' houses, but found that available commercial turbines were too small to do the job. He set about designing a wind turbine larger than any previous one, and by 1939 had obtained early financing from the New England Public Service System. He then engaged the firm of S. Morgan Smith, a manufacturer of hydraulic turbines, to build the unit, which would generate 1.75MW—enough to power a small town. A me-

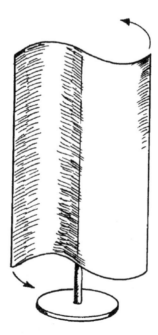

Fig. A.1 Savonius' turbine

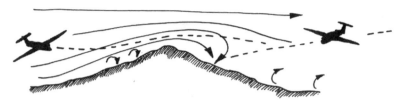

Fig. A.2 Wind flow over a mountain

ticulous engineer, Putnam set out to study the ecological effects of wind, the effects of icing (which would decrease the efficiency of the blades), the type of wind generator that would provide the greatest efficiency (he eliminated the vertical Savonius turbine, which is similar to the ancient Persian machine, because it is less efficient due to the large amount of metal in the swept area (Fig. A.1), and the kind of controls that were to be built into the unit (pitch control vs. cheaper flaps). Finally, he chose the site from among many, considering consistency of wind and favorable topography (Fig. A.2). Restricting himself to New England, he chose as the site the bald top of a mountain in southern Vermont that was christened Grandpa's Knob, since it had been bought from a farmer whose grandfather once owned the mountain. In mountainous areas such as this one, only the ridge crests offer optimal conditions for siting a wind generator, whereas in a flat plain such as the central plains of the United States, no such restrictions would have been necessary (Fig. A.3). Aware of the variability of the wind, Putnam also took into account the need for maintaining power output when the wind blew at less than 7.6 m/s (17 mi/hr). He provided a dammed pond that would feed into a hydro generator on those occasions.

To provide the extensive scientific input needed to realize this project, Putnam assembled a superb team of scientists that included Theodor von Kármán, a leading aerodynamicist, Vannevar Bush, the dean of engineering at MIT, and experts in meteorology, turbine design, and stress analysis. As chief engineer of the project, Putnam chose John B. Wilbur, an eminent professor of civil engineering at MIT. It became Wilbur's job to see the project through the testing phase and onto completion because Putnam spent the war years of 1940-1945 with the Office of Scientific Research and Development in Washington DC, developing amphibious weapons.

The skeleton of the tower took shape in early 1941, in freezing temperatures and numbing wind. Construction of the 52-m (175-ft) diameter turbine proceeded slowly, as each piece of the 250-ton machine had to be carefully hauled up the mountain along a specially built road. As the main girder that was to cradle the generator was being transported, it toppled off and fell into a crevice as the truck turned the last hairpin near the summit. Miraculously undamaged, the girder was dragged out and erected onto the waiting tower.

It took another four months to attach the eight-ton stainless steel blades on the shaft, but finally, on August 29, they were, for the first time, turned by the wind. For the next two months, adjustments were made to the intriguing control systems devised by Putnam to safeguard the operation of the generator. Finally, in the presence of a committed group of men, power from the wind, for the first time, flowed into a utility grid. For the next four years, the engineers carried out a meticulous series of tests to bring the machine to perfection as an automatic generating station. During that time, it survived gales of 51 m/sec (115 mph) and generated power in winds of 31 m/sec (70 mph) without damage except for the loss of a main bearing, which took two years to replace because of the war. All during the testing phase, the unit groaned and creaked, oil seals failed and were replaced, and the stainless-steel skin on the blades cracked and had to be welded but the unit kept working. The data gathered from this test is still used today in the design of modern wind generators.

On March 26, 1945, Wilbur called Putnam with bad news. "Put, we've had an accident. It could be worse. We've lost a blade, but no one is hurt, and the structure is still standing." At 3:10 in the morning, a blade had been ripped from the shaft and thrown 230 m (750 ft) down the mountain. Before the windmill could be shut down, the remaining blade was also damaged as it nicked a tower leg. Von Karman believed that during the two years the unit had been shut down with the blades locked in position, the blades, prevented from rotating and changing pitch, waved like fishing poles, stressing the area around their root. This 'fatigue' stress contributed to its failure.

The turbine was mortally wounded and would never run again. Although the project was abandoned for lack of continuing financing, it was pivotal in proving the feasibility of using large-scale wind turbines in a commercial environment. Putnam never again par-

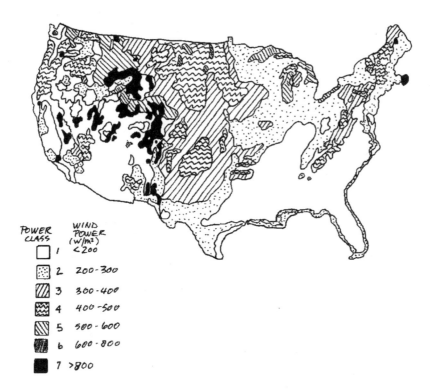

POWER CLASS | WIND POWER (w/m²)
1 | < 200
2 | 200 - 300
3 | 300 - 400
4 | 400 - 500
5 | 500 - 600
6 | 600 - 800
7 | > 800

Fig. A.3 Average wind speeds across the United States

ticipated in the design of a wind turbine, but he wrote a book that detailed every step in the enterprise, providing to others all the scientific information obtained from the years of testing. This served as a guide to the next wave of development.

The oil crisis of the 1970s sparked a renewed interest in wind power. Utility-scale wind farms were developed in California, and isolated wind generators were built in other parts of the country so that by the year 2000, 2600 MW, representing 1% of the nation's need, was supplied by the wind. This is projected to increase to 5% by the year 2020, replacing 60 conventional power plants, with concurrent benefits to the environment (see Chapter 14). By the same time, the European Union expects to generate 20% of its energy from renewable resources including the wind. A 200-turbine wind farm on the coast of Ireland will generate 10% of that country's needs. Denmark expects to generate half of its energy from the wind—it already generates 20%. Germany is currently building 5-MW wind generators with 180-m-high towers and 60-m-long blades, dwarfing any that preceded them.

When Putnam retired to California, he could see the fruits of his dream come true as wind farms powered his new home, but he would undoubtedly be saddened by the fact that his country is so far behind in exploiting this form of non-polluting energy.

B

Appendix B
The Power Of Water

Big whorls have little whorls,
Which feed on their velocity,
And little whorls have lesser whorls,
And so on to viscosity.

Lewis F. Richardson (about 1900), Meteorologist

The earth's water cycle lifts water into the atmosphere through evaporation and then releases it through condensation in the form of rain or snow. This precipitation falling over the land must, because of gravity, descend back to the earth's oceans. In so doing, it releases its *potential energy* and is the engine that gives water its power. The ancient Egyptians recognized this when they built the world's first dam in the Wadi-el Garawi, sometime between 2950 and 2750 BC. Called the Sadd el-Kafara or Dam of the Pagans, it was only 11 m (37 ft) high, but provided water power for an alabaster quarrying operation downstream (as well a providing water for the men and animals who worked there). The dam was poorly built and did not long survive—and Egypt would not see another major dam built for 4700 years. However, in the intervening years, dams were built as reservoirs throughout the rest of the world, although the potential to exploit the power of water did not lead to serious dam building until the eighteenth century.

Near the town of Almendralejo in northern Spain, a 19-m (64-ft) high dam was built in 1747 with the express purpose of turning a waterwheel for a flour mill. A low-level pipe penetrated the wall of the dam to direct water against the paddles (or vanes) that were distributed around the rim of the 'undershot' waterwheel (Fig. B.1), turning it and thus powering the mill. Placed in a fast-moving stream, such waterwheels had been used since the times of the ancient Egyptians to drive millstones to make flour. They were considered so vital to the needs of a region that when the Goths destroyed Rome's mills in their attack of 536 AD, the defending commander, Belisurius, deployed floating mills in the river Tiber (Fig. B.2). In response, the Goths tried, unsuccessfully, to jam these by floating bodies of their dead down the river.

In later times, using a pipe or sluice to lead water to the top of an 'overshot' waterwheel, gravity helped do the work that, in an undershot waterwheel, was provided only by a fast-moving stream (Fig. B.3). John Smeaton, an eighteenth-century English engineer, proved that such a wheel is twice as efficient as the undershot type because both the speed *and* weight of water contribute to driving the wheel. The development of watermills also benefited from

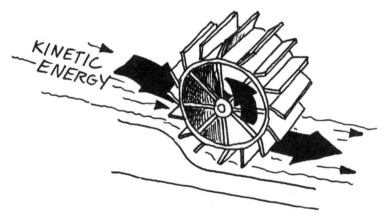

Fig. B.1 Undershot water wheel

Fig. B.2 Roman floating water wheel

a shortage of slave labor. That was the reason announced by the Romans for building a watermill near Arles in southern France in the fourth century. However, little development took place after the collapse of the Roman Empire until the late tenth century. Gradually, every village where water was available had to have a mill, so by the time the first settlers arrived in America, there were 27,000 watermills in England alone.

By the end of the nineteenth century, the power of water had been harnessed globally, as electric energy became the prime source of power.

Fig. B.3 Overshot water wheel

Fig. B.4 Water flow through turbine in a dam

Rivers of Energy

The idea of enclosing a waterwheel and using water under pressure to drive it was first proposed by Jean-Victor Ponçolet in 1826. Many variations and improvements of such pressure turbines were developed in the ensuing years, all with the common characteristic of having high-pressure water directed against vanes around the perimeter of a shaft. To obtain a source of high-pressure water, dams were built across river valleys to impound water and lead it down through a penstock (a tube) to a turbine (Fig. B.4).

One of the world's first such hydroelectric plants was built in 1879 at Niagara Falls, which separates Canada from the United States. The presence of the falls already provided the elevation difference between inlet and outlet levels, so only a diversionary channel needed to be built at the top, to lead water into a penstock pipe that led to a turbine at the bottom. A hydroelectric plant is extremely efficient, as it uses renewable energy, converting 90% of the available energy into electricity. The twentieth century witnessed an explosive growth in development of new hydroelectric facilities.

Of the developable capacity worldwide, one third has already been exploited and provides 20% of the world's electricity. However, damming rivers is not without negative environmental consequences.

Interrupting a river's flow interrupts the natural flushing action of moving water, causing silt to accumulate behind a dam. In time, such siltation can result in decreased effectiveness of the impounded water and may eventually require demolition of the dam. A dam also disrupts fish movement along a river and interferes with migration and spawning patterns of many species of salt-water fish such as salmon and striped bass. Fish ladders and bypass channels that have been built at some dam sites only partially address this problem. As the need for power expands and as the benefits of renewable energy are realized, dam building continues, although with greater caution. One of the most contentious such projects is the Three Gorges Dam and hydroelectric power plant on the Yangtze river in China (see p.124). Initially promoted as a flood-control project, the plant will generate over 18,000 MW, promoting the development of central China. With an ultimate capacity of 22,400 MW, it will be the world's largest hydroelectric project and one of the most critically debated. The reservoir behind the dam will flood over 600 km², displace over a million people, submerge 1300 historical sites, and forever change the breathtaking beauty of the Three Gorges site. Some scientists have also predicted that siltation behind the dam will seriously reduce its power-generating capacity within a few years. Only time will tell if the benefits outweigh the environmental damage.

Pumped Storage

The need for electric power varies throughout the day and with the seasons of the year. Rather than add capacity to meet the need of peak periods of demand, a clever engineer, whose name is lost to us, suggested using the excess electricity generated during periods of low demand to pump water up to an upper reservoir. During peak demand periods, the water descends through a penstock to drive turbines before being released into a lower reservoir. By this means, a better balance is created between demand and supply of electricity.

Tidal Generator

Tides rise and fall depending on the position of the moon[32] in relation to the earth, and vary greatly around the globe. Some locations have large tidal variations: These include the Bay of Fundy, the Severn Estuary, the Cook Inlet in Alaska, the San Jose Gulf in Argentina, the Rance estuary in Brittany, and four locations in Russia. Taking advantage of a 13-m (44-ft) tidal variation, a 240-MW plant (enough to supply a million homes) was built at St. Malo in northwest France into a tidal barrage across the Rance River. This plant is fitted with turbines with reversible blades that are turned by both the incoming and outgoing tide. A tidal turbine is like a revolving door anchored to the seabed, and will operate practically in as little as a three-knot current.

Wave Generator

Two types of generators have been developed to harness the power of waves. The first uses the rocking motion created by a passing wave to activate a generating device in a sealed, floating unit. Called Salter's Ducks, after Dr. Salter of Edinburgh University, they appear like a linked group of ducks bobbing on the waves. The second type of wave generator utilizes the force of water as it hits a coast, especially a rocky one. A vertical tube, open at the top and bottom and containing a turbine, is placed into a rocky coast where waves generally break. As waves rush in, air is pushed out of the tube and then sucked back in as the wave recedes. This causes the air to turn the turbine, generating electricity.

---

32   The gravitational attraction between the moon and Earth causes the tidal variations as the Earth faces and turns away from the moon.

Index

About the Author

Matthys P. Levy is a founding Principal and Chairman Emeritus of Weidlinger Associates, Consulting Engineers. Born in Switzerland and a graduate of the City College of New York, Mr. Levy received his MS and CE degrees from Columbia University. He has taught at Columbia University and Pratt Institute and lectured at universities throughout the world.

Mr. Levy is the recipient of many awards including the ASCE Innovation in Civil Engineering Award and the IASS Tsuboi Award. He has published numerous papers in the field of structures, computer analysis, aesthetics and building systems design, has illustrated two books and is the co-author of *Why Buildings Fall Down, Structural Design in Architecture, Why the Earth Quakes, Earthquake Games and Engineering the City.*

Mr. Levy is a member of the National Academy of Engineering and numerous professional societies. He is a registered Professional Engineer in the US and Eur Ing in Europe; he was also a director of the Salvadori Center that serves youngsters by teaching mathematics and science through motivating hands-on learning about the built environment.

Projects for which he was the principal designer include the Rose Center for Earth and Space at the American Museum of Natural History, the Javits Convention Center and the Marriott Marquis Hotel in New York, the Georgia Dome in Atlanta, the La Plata Stadium in Argentina, the One Financial Center tower in Boston, Banque Bruxelles Lambert in Belgium, the World Bank Headquarters in Washington, DC, and a cable-stayed pedestrian bridge at Rockefeller University, the WW II Museum in New Orleans and the Marine Corps Museum in Alexandria. He is the inventor of the Tenstar Dome structure, a unique tensegrity cable dome used to cover large spaces with minimal obstruction.